ONE

CHINESE

MOON

J. TUZO WILSON

FIRST EDITION 1959

Copyright © J. Tuzo Wilson

Library of Congress catalogue card number
59-14657

Designed by Arnold Rockman

Manufactured in Canada
by
McCorquodale and Blades (Printers) Limited

CONTENTS

Indulgent reader,

be warned that an altogether worthless fellow

who has written no books now essays to be an author.

Only a weakness of understanding

and an excess of conceit

allow him to offer his services as a guide

on a journey along the labyrinthine paths of China,

concerning which

he should be the first to plead ignorance

and to admit but a passing acquaintance.

The result of his labours

would indeed be unworthy of consideration

by your discerning eye and spacious intellect

if it had not been smoothed by the skill

and lightened by the wit

of that fountain of tranquil comprehension

to whom this book is fondly dedicated,

MY NUMBER ONE WIFE

O

=

Jock Wilson is a distinguished scientist. He is Professor of Geophysics in the University of Toronto, President of the International Union of Geodesy and Geophysics, and — just as well, perhaps — an inveterate globe-trotter. No one was surprised that he should have gone to China, and chosen to enter by the most uncomfortable route. His friends get postcards from him from the most inaccessible parts of the world. But what was surprising was that he should have written a book. We expected nothing more than a postcard. But his book, like a good postcard, is personal, cheerful, newsy, and to the point. More important, it admirably describes what a country under Communism looks like to a scientist.

Travellers may have the most conflicting impressions of the political feelings of a country, but not scientists of its science. Scientific endeavour is difficult to simulate. Communists take science, and indeed all learning, seriously. How seriously, and how effectively in China, Professor Wilson tells in this book. Science dries up when politics obtrude. They have not, so far, obtruded in China. Professor Wilson was able to meet freely and on common ground with Chinese scientists. They seem to have liked him, and he obviously liked them.

Attempts were made, though not by scientists, to impress him with the party line. Communists attribute all success, even scientific success, to the virtues of Communism. This kind of nonsense Jock Wilson met with a polite but de-

ix

vastating grin. What did his official guide, the dull and dedicated Mr. Tien, make of this irrepressibly cheerful, insatiably curious, but politically undentable Canadian?

But Professor Wilson was impressed with the technical undertakings and serious scientific purpose of the Chinese. He was able to see, and came to admire, their very noteworthy achievements. But — as probably would most Canadians, to whom large-scale technological projects and successful exploitation of natural resources are no novelty — he attributed these more to the nature of science and to the innate sense of the Chinese people. The good party man, however, must attribute them all to the baleful inspiration of an ideology that is to us, after all, old (and none too clean) hat.

No student of the earth sciences can fail to be impressed by the size of China, and the proportion of the earth's population that it supports. The fate and fortunes of the Chinese people concern us all. We need to know more about them. We need greater contact with them. They have contributed notably to civilization in the past. Their sun is in the ascendant again. If the past is a clue to the future, they will contribute once again.

At all events, it seems sheer folly to behave in the international arena as though they have ceased to exist. Links, as Professor Wilson has shown, can be successfully forged in non-political spheres. Perhaps in the end, and despite the politicians on both sides, such links will prove to be the more enduring.

Go Home Bay
Ontario

W. A. C. H. DOBSON
Head of the Department of
East Asiatic Studies,
University of Toronto.

x

PREFACE

This book describes a journey across China as accurately as I can recall it and in as much detail as I feel the reader can comfortably stomach. It is based upon diaries kept for my own satisfaction, weighted with my own interests since I had little thought at the time of writing this book.

Inasmuch as I had never been to China, had no intimate connections with it, did not plan the journey long in advance, and had little time to read the opinions of others, I believe that I can claim that this account is reasonably impartial. It naturally gives the first impressions of a traveller and not the dicta of an expert. My chief concern was to try to create friendly international relations between scientists.

I should like to thank the National Research Council of Canada for financing some of the expenses of my devious return journey from meetings in Moscow; the Department of External Affairs for aiding my trip across Asia; Mr. David Johnson, Canadian Ambassador in Moscow; Mr. Duncan Wilson, the British Chargé d'Affaires, and Mrs. Wilson; and the members of the diplomatic staffs, for their warm hospitality and unfailing help.

I am grateful to the Academia Sinica for their invitation to be their guest while in China, and to the secretary-general, Dr. Pei Li-shan, and other officials who made the arrangements for my visit; and likewise to the Soviet Intourist and other Russians who facilitated my journey across Siberia. To

Mr. Tien Yu-san I owe my gratitude in a particularly personal way for his skill as an interpreter, his courtesy as a companion, and his diligence as a guide.

I have mentioned the Chinese scientists who made me welcome at the various institutions, but I am particularly grateful to Drs. Chow Kai, Lee Shan-pang, Chang Wen-yu, Chen Tsung-chi, Wang Yung-yu, and Tung Chieh, who bore the brunt of my entertainment and on whose time I trespassed the most. I shall long recall with pleasure my reception in Hong Kong by Dr. and Mrs. L. G. Kilborn.

Immediately upon my return from China I attended the 1958 meeting of the International Council of Scientific Unions in Washington, D.C., as a representative of one of its constituent unions—the International Union of Geodesy and Geophysics. Dr. Lloyd V. Berkner, then president of the I C S U, introduced me to Mr. John Lear, Science Editor of the *Saturday Review,* who was seeking information about science in the People's Republic of China and who persuaded me to write an article which he edited and published in the Science and Humanity section of the *Saturday Review.* Publishers who read that article commissioned this volume. I am grateful to Mr. Lear for having encouraged me to write this book and for allowing me to use some of the *Saturday Review* material in Chapters Nineteen and Twenty.

But this book would probably never have been written if Helen O'Reilly, Gregory Clark, and Laurie M. McKechnie had not goaded me, literally thrusting the microphone of a tape recorder at me and ably editing the result.

For reading the text or answering questions, I am indebted to Mrs. Graham Rowley, Mr. E. R. Hope, and Professors W. A. C. H. Dobson, L. C. Walmsley, and W. H. Watson. Their scholarship has pulled me back from the brink of many pitfalls, but they cannot be held responsible if I have blundered into others.

Mr. Alex Aiken was so kind as to draw the sketch map illustrating my travels. Susan Bradshaw, Sylvia Derenyi, and

Nancy White interpreted my scrawl. The photographs I took myself.

To simplify matters for the reader I have converted currency, weights and measures to their North American equivalents on the basis of forty cents to the yuan and substituted our linear measure for the metric and local systems used in China.

Go Home Bay
Ontario, Canada
July 5, 1959

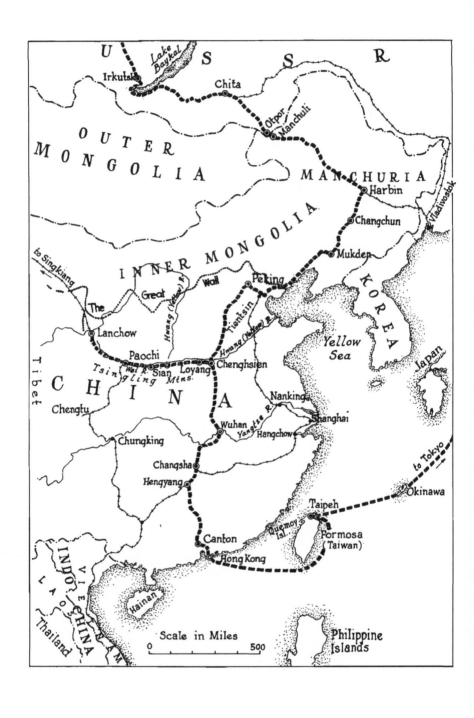

U S S R

Lake Baykal

Irkutsk

Chita

OUTER

Otpor

MONGOLIA

Manchuli

MANCHURIA

Harbin

Changchun

Vladivostok

INNER MONGOLIA

Mukden

KOREA

to Singkiang

Wall

Peking

The

Great

Hwang (Yellow) R.

Tientsin

Japan

Lanchow

Hwang (Yellow) R.

Yellow

Sea

Paochi

Chenghsien

Tsin

Wei R.

Sian

Loyang

gling Mtns.

Tibet

C H I N A

Nanking

Chengtu

Shanghai

Wuhan

Yangtse R.

Chungking

Hangchow

Changsha

to Tokyo

Hengyang

Okinawa

Taipeh

Quemoy

Canton

Isl.

Formosa

Hong Kong

(Taiwan)

INDO

VIETNAM

LAOS

CHINA

Hainan

Thailand

Scale in Miles

0   500

Philippine

Islands

O

Hall of Precious Harmony

On Tuesday, August 26, 1958, as the morning sun rose over the Yellow Sea, it brought dawn to China and flooded through the windows of No. 4 Express of the Russian State Railways on the eighth and last day of its run from Moscow to Peking. It awakened me in Lower Berth 3, Compartment 1, Car 4.

Quickly I got up to watch the golden sunrise.

The train was running along an embankment by the edge of the sea where the last fields give way to salt marshes. Fascinated, I watched fishermen tending their intricate bamboo traps among the reeds. Along an ancient canal, laden barges sailed almost imperceptibly forward. In crowded villages of mud huts, women washed clothes, swept dust and nursed their children. In this peaceful corner of an ancient land, nothing appeared to have changed.

On the train there was an air of expectancy and stir after a monotonous week of incessant travel. I welcomed the thought of arrival in Peking, the excitement of visiting China for the first time, the possibility of a good bath and the hope of a quiet night's sleep.

A few hours later and precisely on time the great express rattled through the environs of the city and stopped in a crowded, modern station.

The idea of going to China had first occurred to me the previous winter when I was appointed by the Canadian Government to be one of six Canadian delegates to the fifth and final meeting of the special committee that organized the International Geophysical Year. That meeting was to be held in Moscow from July 29th to August 8th. To my mind one should never miss a chance to see something new. It seemed to me that it would be more interesting to return to Canada through Asia than to fly back by the way I had come —across western Europe.

Thus my visit to China started propitiously, stemming as it did from a splendid meeting marked by a feeling of triumphant and good-humoured fellowship, natural among men who had managed a great co-operative effort in the midst of a divided world.

The meeting was attended by several hundred scientists from most of the sixty-six countries participating in the International Geophysical Year. Indeed, my whole visit owed so much to the opportunities provided and to the good will generated by the I. G. Y. that I must say something about it.

The International Geophysical Year will long be remembered as an important event in the history of science. The initial conception came from a group of scientists, meeting privately in April, 1950. It was a direct descendant of the First and Second Polar Years of 1887-1888 and 1932-1933, but its scope was much greater. It became the first and only project in which scientists of most of the world's countries have co-operated to make an organized study of the physical nature of the entire earth.

The earth is rapidly filling with people. They are migrating to remoter, less fertile domains in less temperate lands. They will have to increase the productivity of their fields. As a wise farmer before breaking new ground must first ex-

2

plore its possibilities, so must mankind survey the earth to learn more about its physical nature.

The world's geophysicists who undertook this task through the I. G. Y. plans were concerned with weather and climate; they examined water supplies in oceans, rivers, lakes and glaciers; they recorded earthquakes; they organized intensive, world-wide studies of such varied and complex matters as the earth's magnetism, its upper atmosphere and the disturbances caused by aurora, cosmic rays and meteors. These latter do not affect food supplies, but they influence the intricacies of navigation and the clarity of communications—both vital in this age of jets, missiles, television and space travel. The Sputniks and Explorers resulted from their planning.

Above all, man's paths are lighted by the sun and there is little he can do on earth that is not directly affected by its benign radiation; so the I. G. Y. program also included study of the sun's effects upon our earth. Sixty-six nations co-operated to make these studies, not only within their national boundaries but in every remote corner of the globe.

When the final meeting of the organizing committee was held in Moscow, the eighteen-month "year" had already run two-thirds of its course and preliminary results were pouring in. The great problems of organizing this complex international endeavour had largely been solved. It was too late to fill any gaps remaining in the network of scientific stations. Only the exact method of winding up so vast an enterprise and of securing its results for posterity involved any argument and debate. Because the plans had been so carefully laid, the Moscow meeting, to a much greater extent than its predecessors, was freed of administration, and from the floods of information gathered around the world from pole to pole, the scientists were proudly able to present their first conclusions.

Cold scientific facts prompted warm, human thoughts of observers isolated on the most distant and inaccessible islands of the uninhabited oceans; of ships, often storm-tossed, on the seven seas, whose crews were probing the depths far from normal shipping lanes. We realized that on mountain tops scientists had struggled out of their shelters in incessant gales to win the knowledge which we exchanged, and we remembered the muffled figures who, in the polar nights, had faced temperatures as low as 125°F. below zero to discover the nature of the extremities of the earth.

The spectacular launching of the satellites had attracted the attention of the world, but at this Moscow meeting we became fully aware of the dramatic achievements of a host of men who helped make the launchings possible and who had collected the data from the hurtling artificial moons.

The International Geophysical Year was, moreover, a new experiment in international organization. Each of the sixty-six countries was fulfilling its own agreed part in a common plan, with the result that among them there was a lively spirit of competition to perform their separate shares of the task with the utmost excellence. Until the separate pieces of this jigsaw puzzle of scientific data were put together, the picture would be incomplete.

The information sought was not of a nature that nations keep secret. There was an agreement to exchange it freely. So far from holding back, competition was keen to display the greatest accomplishments. That atmosphere bred goodwill.

Perhaps the greatest achievement of the I. G. Y. was this demonstration that the people of the earth can work openly and harmoniously together in a world-wide enterprise. True, it was a simple, basic project, easily divorced alike from politics and emotion, but within its own scope it provided brilliant evidence that men of all nations can work in harmony.

The I. G. Y. may be likened to a tremendous campaign, short-lived and handled as a part-time (and especially as an over-time) job by its administrators. It was never intended

4

that it should become a permanent institution. In this respect it resembled a drive for some great community project, to which citizens contribute their efforts eagerly for a time and then return to their customary occupations.

I sometimes feel that a great injustice is done to the scientists of all nations who participate in these international ventures. The suspicion seems to be all too readily aroused that between the citizens of countries which are politically divided, co-operation cannot be real and hospitality cannot be genuine.

The fact is, I can think of no more homogeneous group in the world than its senior scientists. They avoid discussions of topics extraneous to their own subjects which might engender bad relations. Many of them have met before; they all have acquaintances in common; they read each other's papers, they write each other letters.

Among themselves they debate and argue vigorously, but the divisions between the debaters on scientific subjects are not on national lines. They are anxious to strengthen their reputation for knowledge and can do this only by exhibiting their discoveries. Jealous of their stature as scientists before international gatherings, they strive to subdue the heat of emotions and try to gain respect by the logic rather than by the vigour of their arguments. Naturally, these men are loyal to their native countries and to their convictions and faiths. Debate upon these matters has its proper place, but this is not at scientific meetings. Scientists everywhere are suspicious of those who endeavour to substitute propaganda for learning and logic.

It was under these circumstances that I arrived in Moscow in July, 1958, and it was from these meetings that I planned to continue my journey eastward rather than to return across western Europe.

Before I reached Moscow I had already taken the preliminary steps for travelling home through China. I had consulted the Canadian Department of External Affairs and discovered that, although Canada had not yet officially recognized the government of the People's Republic of China, they had not the slightest objection to my visiting that country if I cared to do so. They recommended me to the good offices of the British Embassy in Peking, but they cautioned me that if I were so foolish or unfortunate as to get myself into trouble, there would be little they could do to help. They advised me that I could get a visa from the Chinese Embassy either in London or in Moscow.

I did not know any Chinese scientists personally (although I had read the work of some and knew their names); therefore, I had written to the secretary-general of the Accademia Sinica in Peking and asked him if I could see something of the geophysical work being done in China. He had replied in cordial fashion, inviting me to be the guest of the Academy while I visited some of their geophysical institutes and university departments. I made similar arrangements to visit Taiwan and Japan on my way back to Canada.

These arrangements were completed in Canada in the spring of 1958, and on the sixteenth of July, I flew from Canada to Europe.

One-way ticket to Siberia

I arrived in Moscow on the evening plane from Helsinki with Prof. W. A. Heiskanen of Finland's Geodetic Commission, and we were greeted at the airport by Prof. V. V. Beloussov, the distinguished Russian geophysicist. They are the two vice-presidents of the International Union of Geodesy and Geophysics.

The first afternoon on which we had a break in our meetings, I went to the Chinese Embassy and presented my passport and my invitation to Peking. It was a moderate-sized building, rather simple by embassy standards. After a little stumbling in sign language, Chinese officials appeared who spoke English and received me courteously. They took some particulars, said that they would look into the matter and asked me to return in a week's time. On my second visit they immediately stamped a Chinese visa into my passport.

Thus armed, I presented myself to the Intourist travel desk in our hotel, but I discovered I had come at the wrong hour. No one was there who could make any arrangements about travelling. The next day I was more fortunate and I was greeted by a ravishingly lovely Uzbek princess who in spite of her origins had not, I quickly realized, the remotest idea about travel in Asia.

"Could I please have a ticket to Peking on the Trans-Siberian Railway?" I asked.

"I'm not really very sure," she replied, "but if you would like to go to the Black Sea or to Leningrad I could certainly arrange to put you on a tour."

This was a little disconcerting, and I felt that I was distracting her attention from the doe-eyed Uzbek prince who was trying unsuccessfully to hold her hand under the desk.

I may say that this was an unusual departure from correct and proper Intourist standards. All the other Intourist personnel I encountered were extremely efficient. Sometimes they were clearly forced by regulations to employ delaying tactics, but this beautiful creature was the only one I ever met who did not seem to have the least idea about her job. She belonged to a "National Minority", as non- Russians in the Soviet Union are called. Perhaps that is why she got away with it. I found her oriental beauty so entrancing that I forgave her.

During the month I spent in Russia, I showed up often at this and other Intourist offices. Whenever I was not busy at meetings or off on a trip, I would drop in and ask, "Where is my ticket to Peking on the Trans-Siberian Railway?" They would reply: "But you would find it so much more comfortable and quicker to take the jet; Peking is only eight hours away, including the stops for refuelling at Omsk and Irkutsk."

Eventually, in desperation I sought Ludmilla's advice. Ludmilla was one of the interpreters we saw every day. I had already learned enough to know she did not ordinarily deal with travel arrangements, for Intourist have a complicated division of labour, but I had also come to know Ludmilla and her inexhaustible willingness to be helpful.

"Ludmilla, I know it is somebody else's job to get tickets, but I have been trying for three weeks to get a berth on the Trans-Siberian Railway with no success. Could you please see what you can do?"

When another day had gone by, she arranged to take me to the head office of Intourist, upstairs around the corner from the National Hotel, to a young man who, Ludmilla said, dealt particularly with Canadian and British visitors. He greeted me warmly in perfect English, and assured me it would be easy to fly by jet aircraft from Moscow to Peking.

I said I appreciated that, but I would prefer to travel by railway or, if that was impossible, then by the slow plane which I understood travelled in easy stages over the old Silk Road to China.

"Why," he asked, "can you possibly want to spend eight uncomfortable nights and days on a train when you can get there in eight hours in one of our magnificent new jet aircraft?"

I explained that I had already been to the Caucasus Mountains for a week-end and back in a TU-104 jet plane and that I did indeed appreciate its wonderful qualities, its luxuries and comforts, but I very much wanted to go by train. I said: "I am a geologist; the land fascinates me. I have long wanted to see the steppes and mountains of Siberia, to cross the great Ob, Yenisei and Lena Rivers and to skirt Lake Baikal. The director of your All-Union Geological Institute in Leningrad has told me that Lake Baikal lies in a graben. I know that you don't know what a graben is, but the director urged me to see that superb graben."

I do not know whether this technical talk and name-dropping had much effect, but he was clearly impressed by one thing—my invitation to China. The letter of invitation, sent me by the Academy of Sciences, was written on official notepaper with the letterhead printed in Chinese, in Russian and in Latin. The letter itself had been typewritten in Chinese ideographs. I found this a most valuable document.

All the Russians could read the letterhead; none knew what the letter meant, but the fact that I possessed it clearly made me a man to be wary of. Neither would *I* have known what it meant had the Chinese not been thoughtful enough to send an English translation along with the original. This I thought unnecessary to show to the Russians.

At last on Friday, August 15, when I came back from my last Russian expedition—a most interesting one, visiting parties prospecting for oil on the Volga plains—he solemnly gave me a ticket on the next day's train for Berth 5 in Car 5, which is universally the Deluxe International Class sleeper, the car right behind the diner on European expresses.

There still seemed to be some difficulty about my passport, but I felt I had at least done well in getting the ticket. According to Intourist regulations, I should have bought it in Toronto and paid for it in roubles, costing 25 cents each; but, when I had endeavoured to do so the previous winter, nobody in North America seemed to have tickets on the Trans-Siberian Railway or to know how to get any. So I had to wait until I reached Moscow.

When the young man finally produced the ticket, I inquired, "Will you take roubles?"

"Of course, it is the currency of the country," he replied.

Excusing myself briefly, I went down to the hotel desk and changed the necessary dollar traveller's cheques into roubles at the cheap, local tourist rate of one rouble for ten cents. I thus procured a ticket for an eight-day railway journey across Asia in a top-class berth for $123.50.

I was due to depart at eight the next evening. In spite of the ticket, I still had no very great expectations of doing so. As is the fashion common on the continent of Europe, my passport was held in the hotel office and as long as it stayed there I could go nowhere. Every two hours I would ask for it and inquire about the exit visa but these requests proved futile. I discovered how watertight the Soviet departments are.

In my despair I turned to the Canadian Embassy. Officials told me that had they made the original arrangements, they might have been able to help. Since I had launched my own arrangements there was no way they could usefully interfere. Once again I discovered how watertight are government departments.

I now noticed that those of my acquaintances among the Russian scientists to whom I had appealed for help had all vanished for the week-end, except two who came to the hotel in the afternoon to bid me a very cordial farewell. They said they could do nothing to influence Intourist arrangements or to expedite my passport.

One of the translators came twice and was most solicitous, even to driving me about on last-minute errands in his own small Moskvich car. In earlier trips to the Caucasus and Leningrad these same people had always seen me off at the station or airport. The fact that they clearly did not intend to do so today, plus their great show of cordiality, strongly suggested to me that they suspected that I would not be allowed to take the railway, and wished to avoid becoming involved in my possible embarrassment.

Sure enough, after further postponements at 4, 5 and 6 o'clock in the afternoon, at 7 o'clock a more senior Intourist official, a rather brusque young man, appeared and said: "Your papers are not in order. You cannot take the train."

He explained that the long, eight-day journey by train would extend my visit to thirty-two days in Russia before I crossed the border. Therefore, I would need an extension to my tourist visa. It was valid for only one month—not thirty-two days. I would also need a special exit visa to leave the country at Otpor on the Chinese frontier.

"All this could be avoided if you would take the jet plane, since you could then clear immigration in Moscow within the time limit of your present visa," said the brisk young man.

This was the news I had rather come to expect, but it greatly distressed the little Intourist girl with whom I was dealing at that time. It was only an hour until train time. She very kindly volunteered to run to the railroad station to cancel my ticket and save my money.

11

It was Saturday evening; as in less enlightened parts of the world, nothing could be done until Monday.

I hated to admit defeat or to show myself to my friends. I spent the week-end sleeping, walking, writing letters and listening to Pol Robeson (as the Russians spell it) on the radio. It was evident that he was extremely popular in the U.S.S.R., which he was visiting at that time.

On Sunday evening I walked over to the Metropole Hotel for dinner because it was the only one of the seven Intourist hotels in Moscow at which I had not yet had a full meal. By chance I saw the great American Negro singer pass through the dining room on his way out of the hotel by the side door.

I shared a table with an American doctor who told me that he was inspecting public-health services in Russia. He said that earlier in the day Robeson had been practically mobbed by admirers when he left by the front door. He went on to say that he had been at the concert which I had heard on the radio. He considered it a good thing that Paul Robeson had been allowed to come to Russia.

He was also impressed by the tremendous wealth of Russian resources and the relative poverty of the people. He said to me:

"The Russians remind me of a farmer who lives simply while he buys tractors with which to work his rich farm."

At 9 o'clock Monday morning I was waiting beside the Intourist desk when it opened. I waved my Chinese letter and said:

"I think your jets are wonderful, but I would still like to go by train."

"Come back later," they said.

Rather to my amazement, when I came back at noon I was given my passport complete with an extension of my tourist visa and permission to leave Russia by train at the Manchurian border.

The same little Intourist girl in her same little green sweater (I say little because she wasn't very tall, though I suspect that she really was a married woman of about thirty-five) went off to get me a ticket. It took her the whole afternoon, but at 6:30 p.m. she triumphantly reappeared and explained to me in French (which was her only foreign language) that she could only get me a first-class sleeper, as the international class was sold out. She explained she had trouble procuring even this less desirable accommodation.

Looking back on my little troubles in Russia, I find it hard to decide to what extent the difficulties were deliberate and to what extent they arose from bureaucratic inefficiency. It would be easy for anyone used to the freedom of western travel to blame it all on malicious obstructionism, but the Soviets last year admitted and welcomed thousands of tourists, far more than ever before. I think that the answer lies simply in the complications which arose from this expansion, from the cumbersome system and from my being a deviationist.

The approved deportment of a Western tourist is to confine his visit to one month, to take organized tours to orthodox places, and to depart by jet plane. In the view of Intourist this would give the most favourable impression of the country. All arrangements were geared to such schedules. I not only wanted to stay longer but to travel in an unorthodox fashion. This was not forbidden; it was merely more complicated. It could not be handled in a routine manner; someone had to make a decision. I can see my request travelling higher and higher up the chain of command. I can visualize the exasperated questioning into the reasons why this obstinate foreigner wished to travel in so irregular and uncomfortable a fashion.

One must remember that, with the state in control of everything and with no competitors, there is a lack of initiative and a fear of making any unusual decisions. I had noticed this same lack of initiative in young Hungarian refugees who came to me in Canada as students. Although they were opposed to the Communist system, they had been trained under it and bore its stamp. They always waited to be told.

Last summer in Russia the situation was made worse by the great increase in tourist traffic, with the consequent increase in the number of inexperienced employees in government agencies, and the difficulty of procuring any accommodation at all.

Trains, planes, hotels and theatres were nearly always full. To obtain a place for a tourist very often involved finding some unfortunate Russian sufficiently unimportant to be safely bumped at the last minute.

At any rate my way was clear, and I started my dash for the train. It was 6:40 p.m. when I rushed up to my room on the twentieth floor as fast as a rather overworked elevator could take me.

"Dve tysyachi tri (2003)," I said to the woman on duty on the floor. She knew me quite well by sight and started a tirade in Russian as she gave me the room key. The torrent increased to a crescendo a moment later, when I emerged with my bags. Although I could not understand her words, I could tell she wanted some money and that she was quite determined to get it. She accompanied me down in the elevator and into the Intourist office, scolding and expostulating the whole way.

I had paid for my room before I left Toronto, but in the course of my travels in and out of Moscow, the piece of paper which proved this essential fact had been left behind at another hotel and had not reached the Ukrainia Hotel. She was therefore quite determined to extract 88 roubles from me for the past three nights' lodging.

Having no time to spare, I paid.

While the unsympathetic might jest that I was being gypped, I do not think so in the least. It was another example of the inevitable difficulties which befall the unhappy traveller who departs from the well-worn Intourist routes. I certainly invited difficulties with my many comings and goings

14

and my curiosity to stay in as many hotels and to see as many things as possible.

It was not until 7 p.m., just an hour before train time, that I was finally free to leave the hotel. The young Intourist woman in the green sweater offered to accompany me, but I refused her kindly. She had been to great trouble already on my account and after a month in Russia, I felt that I should by now be able to get a taxi by myself to take me to the right one of Moscow's eleven railway stations.

Half an hour later I was disgorged into a mob of people seething in front of the Yaroslavl Station. Picking up my suitcase, my bulging brief case, my roll of maps and my coat, I shouldered my way into the crowd.

The great waiting halls inside the building were packed with people. Most did not look as though they were expecting to catch a train for days, and they had camped for the night. Children slept on rugs on the floor; old people dozed on benches. From string bags and paper packages, one could see protruding an abundant supply of bread and cucumbers, of cabbage and smoked meat, water melons and bottles of sour cream. Anxiously, I picked my way among the sleeping children, between the bundles of quilts and over piles of packages, and I pushed through throngs of people until I reached the gates to the trains.

At the entrances to the platforms I rushed up and down looking in vain for the departure sign of a train leaving at 8 o'clock and in any way resembling an express for Peking. All appeared to be local trains. I began to worry whether I had come to the right station. One or two uniformed men hurried past who looked as if they might be porters. I tried to stop them, but they were much too busy. No one could understand me. I do not speak Russian. In desperation, I began to shout "Peking" at the passers-by and soon one of them was kind enough to point to an archway which I had not noticed. Beyond it lay the separate express platforms.

I soon found the crack train I wanted. A gate-woman

punched my ticket and with a feeling of infinite relief, I squeezed through the turnstile on to the express platform and into an atmosphere which was in every respect calmer, less hurried and full of greater promise than the crowded station behind.

Stretching far down the platform before me was No. 4 Express of the Russian State Railway System. On every one of its fine green sleepers was a board with the reassuring words that even I could understand, "МОСКВА-ПЕКИН."

At a more leisurely pace I walked down the train along a platform full of officers in uniform, their wives and sweethearts, Russian men of affairs and excited Chinese students. At the door of every sleeper stood two porters, smart in their white jackets, caps and white gloves. Many of the women were well, if simply, dressed, and carried big bunches of garden flowers—gladiolii, phlox and roses.

Many were the embraces and handshakes and great was the laughter and gaiety. For the Russians, the departure to Siberia or to China did not appear to be an exile but rather a great adventure, gaily undertaken, perhaps one involving promotion or increased responsibility. The animated Chinese were going home.

At Car No. 4, the porter looked at my ticket, seized my bags, hustled me unceremoniously into the first compartment and pointed to Upper Berth 3.

Never have I seen anyone more surprised than the middle-aged Russian civilian, his wife, and their fifteen-year-old son, who were the other occupants of the compartment.

"*Spasibo* (thanks)," I said to the porter.

"*Dobree vecher* (good evening)," I said hopefully as I extended my hand to my fellow-travellers, as the saying goes. Then pointing to myself, I said,

"Wilson, *Kanadski*."

There, as far as I was concerned, the conversation ceased. I had exhausted my supply of Russian.

To hide my relief at catching the train, and to give the

Khimishes, as they said their name was, a chance to recover from their obvious surprise, I stepped out again on the platform to catch my breath.

Eight nights in a Russian sleeper

I went back into my compartment. It was like many other European sleepers. On either side between the door and the window stretched a long seat, comfortable enough by day, but at night a rather narrow bed. Above these seats were upper berths which I never had occasion to try because Khimish generously insisted that I should occupy the lower berth which faced forward. He showed me how the seats lifted up, revealing chests below in which to stow baggage.

The porter came and made up the beds with a thin palliasse, two sheets, one heavy blanket, a pillow and pillowcase. Throughout the journey the porter did nothing further about the berths apart from changing the sheets from time to time. It remained for the individual traveller to decide how he would most comfortably travel. I found it most convenient to leave the bed made up and lounge on it through the day as we rolled across the continent.

The compartment was clean and new; its light blue walls contrasted pleasantly with the brown woodwork. The outside window was large and it was washed down from time to time. In front of it and between the seats was a small table upon which stood a desk lamp with Victorian fringes on the shade. On this table Mrs. Khimish had arranged a magnificent bunch of gladioli, phlox and asters. As the journey passed, an accumulation of bottles of jam, pots of meat, jars of pickles and sour cream, boxes of biscuits and meringues grew up around them. This was the larder which supplied the Khimishes with all their meals except a big dinner which they took in the diner every afternoon at three.

I had thought that I was travelling light, with my suitcase, briefcase and roll of maps, but they had rather less. This they quickly stowed away, except for a big string bag of apples from the Crimea which they hung on a peg and which, with the flowers, gave the compartment a pleasant scent.

Having heard awful stories about the stuffiness of Russian trains, I seized the first opportunity when they were out to open the ventilator in the roof. I need not have worried, for they had no fear of fresh air and not infrequently opened the window, which blew a gale directly on my seat. Even the wire screen on the outside failed to keep a considerable amount of soot from entering once we had changed from electric to steam locomotives at Alexandrov Station 60 miles out of Moscow.

The Khimishes were charming people and though I could scarcely communicate with them, I felt very fortunate in having such agreeable and sensible companions. They could not have been more interested in my welfare or more kind. That evening and on succeeding days, we wrote each other's names and exchanged addresses. They corrected the Russian lessons I was beginning to write and set me new exercises.

He was a strong, dark man of medium height, resolute and good-looking. I gathered that he was an engineer from Omsk and that he was returning home with his family from a summer holiday in the Crimea. Certainly they were all sunburned, relaxed and happy. He had the firm, good-

19

humoured manner about him of one accustomed to command. His wife was plump and jolly. She seemed as sensible as he and conveyed to me that she also was employed as an engineer. Their only child was the boy, Victor, age fifteen. A studious type, he read a great deal and played chess.

In all the excitement I had not had anything to eat since lunch. As the train left the outskirts of Moscow and entered the Russian forest, I felt hungry and went back to the diner.

I thought that going to bed would present a problem, but while I was out Mrs. Khimish got into the upper berth which I should have occupied. On my return her husband indicated by pointing that I should sleep in the lower berth directly beneath it. It was not chivalrous, but it was sensible, for then I could go to bed and get up first without any embarrassment. I was sorry when they got off at Omsk.

They were succeeded by a young air-force officer who also had a wife and son. Can you imagine any worse fate for a young Communist career officer than to be cooped up with a capitalist? It was just as though a young American officer were forced to cross the continent in the same compartment as a Communist professor who could speak no English. He had no fear of me, but he feared for his reputation.

I felt that the Russian reacted in exactly the way that a Texan would have. The fact that he and his wife were tall, slim and blond, and looked as Texans should, heightened the analogy.

Being more chivalrous, but less practical than the engineer, he arranged for his wife to sleep in the lower berth opposite to me. One of us had to put his head under the sheets for the other to get into bed. He was probably annoyed with me for continuing to occupy the lower to which he was properly entitled. He observed the conventional courtesies, but never became at all friendly.

There are two ways of looking at the Trans-Siberian Railway, either as a mode of transportation or as a way of life. In either sense it is in a class by itself.

It is the longest railway in the world, the double-tracked economic backbone of the U.S.S.R., stretching 9,302 km. (6,000 miles) from Moscow to Vladivostok. I travelled over most of it before turning south into Manchuria at Chita and thence over the lines of the Chinese Eastern Railway and the South Manchurian Railway to Peking and on to Hong Kong, thus completing a rail journey of perhaps 8,000 miles from Leningrad right across Eurasia.

The gauge in Russia is five feet—three and a half inches broader than ours—so the trains are at least as big as ours. There are plenty of them. Most of the hauling is done by steam locomotives with five driving wheels to a side. One hauls the expresses; two engines double-head the freights. It would do any railroader's heart good in this age of diesels to hear the long moan of the steam whistles at night over the steppes, to listen to the heavy freights panting up the steep grades around Lake Baikal and the Tienshan Mountains and at the divisional points to see twenty of these big loco-motives standing by, with steam up, tooting to one another. For their veritable conversations the drivers use high-pitch-ed compressed-air whistles and not steam.

Traffic is heavy. The main part of the line carries as many as five trains an hour each way—for I noticed that we passed as many as one every five minutes going in the opposite direction.

As far east as Omsk there is a network of lines across Rus-sia with at least three main routes. Our express followed the most northern, but from Omsk on to the Pacific there is only one line, double-tracked, with electric block signals and excellently maintained. It is perhaps indicative of the heavy traffic that the first part of this section is being electrified be-tween Omsk to Taiga (the junction for Tomsk).

All the way, work is in progress building more sidings, putting down heavier rails and replacing sleepers. All level-crossings have gates. There are few of these, for in Siberia there are few automobile roads, only wagon tracks. All sid-

21

ings have switch tenders—usually young women who hold out furled yellow flags or yellow lights for every train. At stations the station-master in a red cap stands on the platform and holds up his wand as the trains go by.

It is a good railway and well run, but as with most railways, (and most of Russia), little has been done to beautify it. Many of the smaller stations where we stopped were little more than freight yards, and even the main stations were remarkable rather for the volume of their freight than their passenger traffic. Invariably, there were several freights, with steam up, on the sidings waiting to follow our express.

The porters often let us get down and walk back and forth between our train and the waiting freights. These were usually made up of boxcars, oil tankers and flat-cars loaded with timber. If there was also a coal car, so much the better ... then one of the porters would climb on top to fill a pail with coal to heat the samovar in our sleeper.

Even main stations were rather dishevelled places; the finer buildings, in ornamental brick, clearly dated from the construction of the railway in 1900. The asphalt-paved platforms were crowded with people. The cars stood high off the ground and a surprising number of people were continually passing beneath them—railway men and women, switch tenders crossing from one line to the next, repairmen taking off and adjusting the belts which drove the light generators on each car, passengers looking for a train, and, always, small boys. It was a sign of the tough, practical and happy-go-lucky attitude of Russians towards life that they would pass under the trains instead of going around them.

Along the main platform there were habitually many stalls and stands, tended by kerchiefed women eager for business. They stood behind long counters selling pickled cucumbers and sunflower heads, boiled eggs, bottles of sour cream, loaves of bread, water-melons, carrots and cabbages. Other stalls had newspapers, books, papers and magazines, ice cream on sticks, and soft drinks. These latter were measured out like chemicals from a tall glass graduate with a tap at the bottom and were sold by tenths of a litre.

In all the way I only saw three copies of papers in Western languages: a copy of *Women's World,* one of *L'Union Soviétique,* and a brochure from East Germany.

From the door of the diner a blonde waitress, assisted by the train's general factotum, could generally be seen doing a thriving business selling delicacies to the locals. A crowd of dirty threadbare small boys and railroad workers always gathered to buy as soon as the train stopped. In her basket the waitress had chocolate biscuits, individually wrapped with a paper showing a bear and three cubs in the forest, and caramels with a most incongruous red crayfish portrayed on the outside wrapper. She had big, fancy boxes full of meringues, flavoured in an exotic oriental way that I could not identify, and, most popular of all, bottles of beer and vodka.

At each stop the porters put on their jackets and white caps and stood around the doors in pairs. They knew all their passengers by now and made sure that they all got back on board in plenty of time.

The idea that Russian trains start without warning arises only because at first one does not understand the signals: the three-minute warning message over the loudspeakers, two toots from the locomotive, the raising of hand lights and the blowing of whistles by the guard up forward, the signal light ahead turning green and the final single toot as the locomotive starts.

At night the porter would take a good look with his lamp under the train before it started. I do not know whether it was to insure that no one still lingered there by mistake or if it was to see that no hitch-hiker rode the rods. One evening a well-dressed young woman tried to get on our sleeping-car without a ticket and was quite put out when the porter stopped her. Once on the outside of a moving express I saw a man standing between two carriages in a most unlikely place for a railroader and certainly no place for a paying passenger.

23

In the vast spaces of Siberia, which we entered thirty-six hours out of Moscow, there was little evidence of local trains. There were crack expresses like ours; other trains made up of old, hard-class sleepers and czarist coaches. Nearly all seemed to be coming from and going to far places—Vladivostok, Peking, Otpor (the Manchurian border crossing), Irkutsk, Novosibirsk, Omsk, and Kharkov—but the signs indicated that fully half of them began or ended their journey in Moscow (or мockва, as the Russians write it.)

Siberia still seems to be a rather empty place with only an occasional large city, set on the shores of some mighty river. At such places one could see the chimneys of factories and rows of small wooden houses. At night the lights along the rivers were beautifully mirrored in the water as we rattled across the great steel bridges and passed the river barges at the docks.

One could not really see a great deal and we had no opportunity or time to leave the stations. For the most part the journey across Siberia reminded me of shuttling back and forth two or three times across the most northerly line in Canada from Edmonton to Hudson Bay.

On all sleeper trains there were three classes: "International," corresponding to Canadian pullman compartments; "Soft," corresponding to tourist class on our trans-continentals; and "Hard," like the colonist cars on which I used to travel to the mines in Northern Ontario as a student thirty years ago. The European's love of privacy leads all berths to be arranged in compartments with respectively two, four and either six or nine berths in each, according to the class.

I had paid for "International" but, for lack of space, got a ticket for an upper in a "Soft" compartment and no refund. Concerning this, however, I did not complain for three reasons: first, no one on the train could have understood me; second, I came to enjoy the glimpses of Russian life which I got travelling in the more friendly and crowded "Soft" class; and third, as I have explained, I had managed to pay for my ticket in cheap tourist roubles, purchased at ten to the dollar. This was a piece of chance good fortune, but it is not a practice that I would recommend to travellers interested in reaching their objectives. In general, the best way to be sure of getting anywhere in Russia is to make arrangements well in advance.

There was no smoking-room or club car—nowhere to go but one's berth, the dining-room, or the corridor. Most of the Russians rather sensibly wore pyjamas or house-coats over their underwear all day and slept most of the time. They read much more than we do and they played chess and sometimes cards. To help matters, the corridor in each sleeping-car had six or eight folding seats along it. These were arranged in pairs under the lights with a folding shelf between them to hold dominoes or even, with care and with the help of a piece of string, a chessboard.

There was a loudspeaker playing Moscow Radio in every room. It had a switch and volume control, but it was generally on because I liked the excellent music and couldn't understand the rest. Occasionally I wondered what was happening in Jordan and Lebanon, but there was no way of finding out.

The children, like all Russian children, were well-behaved, merry and playful in a very quiet, obedient way, so that one would never realize that there were half a dozen of them on the car.

Two nice young porters vacuum-cleaned and washed the car twice a day. They had a room at one end in which the one off duty rode in a top bunk with his head by the door. He could just reach out and pull the braids of the two teenage girls in the car when they passed his way, which they did quite often, as they seemed to enjoy this fun as much as he did.

Tea could be had any time for the asking. It was served clear and hot in a glass tumbler with a metal holder. Accompanying it on the tray was a sugar basin holding huge sugar lumps, each about half the size of a small box of matches and individually wrapped.

The experienced travellers had mugs and, to shave, got hot water from the samovar in the corridor. I had no mug and so had to use cold water. At least there was plenty of that for at every main station a crew of girls climbed on the roof and filled the water tanks with a hose. There was, however, a shortage of soap and disinfectants. Passengers with electric razors had to be more ingenious. In one of the large stations one man even rushed up to the wicket where they sold postage stamps and pushed the electric cord of his shaver to the girl inside. She plugged it in and while he quickly shaved beside the wicket, a queue of people buying stamps passed understandingly around him.

In the centre of the train was the restaurant car, to which most of the passengers went for dinner in the middle of the afternoon. The rest of the day the only customers were plutocrats and bachelors without the foresight to bring food for their light meals. Temporarily I belonged to both classes.

For no apparent reason the car was divided by a glass screen into two rooms, giving it a cozy effect greatly heightened by the two sets of curtains in the windows—white ones and dark blue ones. Bowls of apples, big bunches of phlox and gladiolii brightened every table until they died and were not replaced. The buffet at the end of the car was covered with bottles of wine, boxes of candies and meringues, to which the Russians are much addicted.

The whole resembled a country inn rather than our crisp and utilitarian diners and was nicely suited to the casual dress and habits of the travellers. The general inn-like appearance was heightened by the steward who stood beside a large cash register at the kitchen end in front of the buffet. With a round, red face, and a brown suit and maroon shirt, he looked like an inn-keeper, and surveyed the scene with peaceful satisfaction unbroken by any attempt at work of any sort. Even the cash was collected by the waitresses who

26

added up the cost of a meal with an abacus and pocketed the handful of paper roubles.

The food was good and abundant for the price, and not nearly so exotic as some I had tasted in Russia. Here were no partridges with strawberry jam, no coffee with lemon, no tea with wild cherry syrup.

There was always tea and black bread and white, so good that it was a meal in itself. You could buy fifty grams of caviar or butter if you wanted to. For breakfast, there were omelettes and coffee; for lunch, a steak or chop, fresh tomato or cucumber salad, fried potatoes and rice, canned plums— all washed down with excellent Moscow beer or mineral water bottled in the Caucasus Mountains.

For supper I usually had soup and cheese. There are three principal types of Russian soup, graded in price but all excellent: shchi, of cabbage and meat; borshch, of cabbage, beetroot, tomato, meat and sour cream; and solyanka, of gherkins, olives and meat or fish. The meals were à la carte, but cost no more than those in Moscow hotels. I do not think that I ever paid as much as twenty roubles ($2.00) even when I started with red caviar. But a big dish of black caviar was thirteen roubles. Spread on well-buttered bread, with a bottle of beer, it was a meal in itself.

There were about four waitresses who seemed to be on the job from early morning until late at night. The one at whose table I normally sat was indefatigable in a nice leisurely way. One evening when I came at an odd hour, she asked me to wait till she finished her supper. This was very sensible of her because she knew that I had nothing to do anyway that week!

I can excuse my dilatory appearance in the diner because it became so increasingly difficult to know when to eat or when to get up or go to bed. Regardless of the time zones clicking past the windows, the train, like the timetables, continued serenely on its way by Moscow time. Before we reached the Chinese border and changed the clocks we were

six hours ahead of the local time and getting up for breakfast at 2 a.m.

My waitress was both pleasing and obliging. She would
amiably turn up the overhang of the cloth on the table
to cover a particularly messy place. She added to the inn-like
appearance of the car by looking rather like an English bar-
maid in a white blouse, dark skirt, apron and lace cap. She
was also either very stout or very pregnant. I suspect the latter for she was well-married, having wedding rings on both
third fingers. Russian women, incidentally, are casual about
which hand and which finger they wear their wedding rings
on or whether they wear them at all. I hope that she got
back to Moscow in time to have her baby. It was a long way
and the trains really hadn't room for any more people.

At breakfast I might say to her: *"Dobroye utro.* (Good
morning.) *Pozhalste, kofe, omelette, khleb, maslo, mineral'-
naya voda.* (Please, coffee, omelette, bread, butter and mineral water.)"

At the end of the meal I might rise to: *"Sem'nadtsat'
rublei? Spasibo.* (Seventeen roubles? Thanks.)"

For the whole week this was the approximate limit of my
conversations. With the Khimishes it was about the same
except that sometimes we would get out an English-Russian
phrase book which one of the Soviet scientists had been kind
enough to give me.

"Vahsha professiya?" Khimish would say, pointing to
the English translation. "What is your trade (profession)?"

Hopefully I would look down the list that followed for
some appropriate answer.

"I am a carpenter
     joiner
     locksmith
     docker
     bricklayer
     milling machine operator, etc."

"I am a professor," I would say, hopefully in English.

"*Ya Kamooneest,*" he would say, pointing at "I am a Communist."

The other choices open to me included:

"I belong to the British-Soviet Friendship Society
          a religious society
          a sports club
I am a Labour man
        Liberal
        Conservative
        Democrat
        Republican."

It was a slow way to make small talk and hard to convey one's precise feelings. Occasional smiles seemed more satisfactory.

No account of the train would be complete without describing the general factotum. He had the appearance of a benevolent middle-aged mouse with a smiling, pointed face and the protruding eyes of the short-sighted. On his head he wore a black cloth cap, on his nose black-rimmed spectacles and over all a dark grey dust coat.

The exact nature of his duties escaped me. Perhaps he was the baggage man; but he seemed to be a sort of trouble-shooter. From early morning until late at night, he might be seen perambulating up and down the train. If there was too much baggage for the women to unload from the baggage car, he was there; if there were empty beer bottles to be taken off at Omsk at 1 a.m., he was there—probably he saw to it that some refills were put on board too. If the waitresses all had vanished except for my plump one, he would take their place, and I am sure that it was kindness that made him put his thumb in the soup to make sure it wasn't too hot.

The porters looked after the passengers on each car and collected their tickets. There were no day coaches and no conductors or sleeping-car conductors, nor any apparent need for them. A woman at the back of the train with flags and lamps acted as guard. She wouldn't let me ride on the back platform, so I rarely saw her. The running crew changed with the locomotives, but the rest of the train crew went right across Siberia, sleeping in a special car forward. At the border, only nine sleeping-cars and their porters crossed into China.

The passengers could be divided into three types: the odds and sods, the international class and the regulars. The first would appear from nowhere, typically have a beer for breakfast, buy a bottle of vodka and vanish. One saw little of the international class. They were the *élite* and could be seen briefly as they passed to and from their private apartments, where many of them must have had their meals. The regulars looked like junior officers and middle-aged officials of the party or government. Most of them were accompanied by their families. When the train stopped, they stood outside in little knots by the steps and watched "that mad Canadian" walking up and down. I never seemed able to get enough exercise.

Their wives, in print dresses or house coats, seemed to sleep most of the time, when not occupied in washing or feeding their children and keeping house in the cramped space of the compartments, a matter at which they seemed adept. This same efficiency no doubt accounted for the excellent behaviour of all their children. Very charming, gay little boys and girls they were, but astonishingly subdued, able to play happily by the hour at cat's cradle with a loop of string or to amuse themselves with endless games of chess.

After a busy year, and a month of meetings and visits in Russia, I was glad to rest while the train rolled endlessly on across the infertile plains. My picture of the journey is a blurred one, but peaceful and happy.

During the first full day, crossing European Russia, the farms were prosperous, but on the second day when we had entered Asia, there were only forest and clearings. For the most part the trees were firs, poplars and birches and the fields had oats, potatoes and hay. The next day the view was much the same except that the forest was thinner——there were no firs and no oats. The next day the countryside was similar but poorer still. Now there were no poplars or potatoes—nothing but tundra with clumps of dwarf birch trees. That was near Taiga, the junction for Tomsk.

It was too bad for all the good stories concerning Omsk and Tomsk, that Tomsk is in fact not on the Trans-Siberian line at all, but on a branch line fifty miles north of the bleakest part of the run. It must be quite a place !

On the fifth day we got to Lake Baikal and both climate and scenery improved. The firs were back and we saw hilly country for the first time since we had passed the Urals during the second night.

On the sixth morning we climbed up the Manchurian border. It was raining over the wind-swept, grassy steppes. The local girls on the platforms wore padded coats and woollen scarves. The tops of the few patches of potatoes were frozen, although it was still August.

It was a bleak and empty land. Each fine evening I watched the swelling crescent of the new moon riding higher over the steppes and flickering in and out behind the silhouetted tops of the pine forest.

Suddenly I realized that the waxing and waning of this new moon would light my whole journey across China. It was to be my *One Chinese Moon*.

Crossing the border

As the train climbed slowly towards the Manchurian border, I realized that on such high places as these had history been made. Stretching in grand sweeps, unbroken by a single tree or conspicuous building, these rolling grasslands are part of that backbone of Asia which nurtured the brave and turbulent horsemen who repeatedly ravaged the wealthy lowlands of China, of Persia and of Europe. Hence came Genghis Khan, the Mongols, the Manchus and many others who had stirred and moulded the history of half the world.

Under brooding skies this vast, empty, featureless land seemed to me to typify a border, for it resembled the Cheviots of English and Scottish history. The only sign of activity was an occasional solitary peasant reaping the lusher patches of hay or herding flocks of cattle around an isolated hamlet of huts. Water was clearly scarce, for a couple of drilling machines were at work. The peaceful rural scene was appropriate for the boundary of two great neighbours who proclaim their love of peace so volubly.

I was, therefore, somewhat surprised when all the army officers got off the train fifty miles from the border, especially since there was nothing to be seen except a few dozen wretched wooden houses, some new, two-storey apartments, and a few cars.

At the next stop the air force officer and his family who had shared my compartment got off with his colleagues. I was equally perplexed—no air field, nothing there but a few huts. As an old army type, however, I was not surprised to notice that the air force had a new station even if it was only a railway station.

At one of the last stops before crossing the border the sun came out for the first time in days. I was trying to get a photo of a picturesque group of trainmen beside a locomotive when a Soviet security officer who had joined the train, came up and addressed to me a flow of Russian in unmistakably stern if not threatening tones. Plainly, he was accusing me of photographing military secrets; I assume, the numbers painted on rolling stock!

As he spoke only Russian, I had a good excuse to play the innocent and I started a quite separate conversation in my most suave English. How long we would have continued these two independent monologues in the doorway of the train is problematical for at that moment the oldest Chinese on the train arrived.

Earlier I had realized that my border crossing might be facilitated if it were known that I was about to become an honoured guest of the People's Republic to the south, so at lunch I had shown my invitation, typewritten in Chinese, to a Chinese student sitting next to me. As I had hoped, he had been properly impressed and had immediately and enthusiastically passed the letter to every Chinese in the dining car.

One of these was the elderly Chinese gentleman who took up the battle on my behalf, and in voluble Russian, gave the security man a proper dressing-down. The situation did not seem to demand my attention longer. At the first pause I shook them both warmly by the hand, said *"Spasibo"* several times and departed to put my camera away for a while.

It is an advantage in some ways to be a poor linguist when travelling !

At Otpor we piled off the train for a three-hour wait. Otpor consists only of a small railway yard, a big station well-encrusted with propaganda, and a fenced compound for the workers' houses. A number of armed guards (three security police with tommy guns and eight military police with side arms) appeared and disappeared again, having made their existence known.

Suddenly to my horror I saw the train steam off taking my camera and all my luggage with it. But it went only a little way to a special siding where I was amazed to see a gang of men begin jacking up the cars to fit them with wheel trucks of standard gauge instead of the Russian broad gauge, $3\frac{1}{2}$ inches wider. I regretted that large signs and a policeman prevented my going to watch this interesting manoeuvre more closely.

I went into the station to struggle with passports, health certificates and money exchange. It was a long and complicated business, as only one man in the station had even the slightest smattering of English, and it was done very thoroughly, requiring the filling out of several forms.

I was interested to find that I could not change my remaining roubles into Chinese money, nor was I allowed to take them with me. I had to buy Russian traveller's cheques with them.

This done, to the accompaniment of shouted propaganda and blaring martial music incessantly pouring from loudspeakers, I relieved my boredom by happily sending postcards inscribed in scarlet Russian script "Glory to the Communist Party" to my most conspicuously capitalist friends at home. I later heard of the consternation of a Hungarian charwoman who had recognized one of these adorning the office of a Bay Street baron—a pillar of the mining industry in Canada.

I now noticed some of the Chinese casually walking past

34

the warning signs to watch the wheel-changing which had so intrigued me earlier. I joined them. No one stopped us as presumably the officials had completed their inspection of the train.

At the border, where the width between the two rails changes, the Soviets, instead of moving passengers to another train, have provided a special siding where powerful jacks (four for each car) lift the sleepers bodily, leaving the wide bogies and wheels behind. These are pushed out of the way; then a crane swings sets of narrower-gauge bogies into position. On to these the cars are carefully lowered. The same train rolls out on the rails of a narrower track! Ingenious, and no doubt it is good for prestige to be able to run the cars of the U.S.S.R. State Railway right into Peking.

We got back on board and the train crept forward in the dark towards the actual border. It was brilliantly flood-lit. Soldiers with rifles and fixed bayonets were on guard. The last Russian I saw was in a pit beside the track. He was gazing up at the underside of the cars with a floodlight to make sure no one was riding the rods out of the workers' paradise.

In the distance through the open window one could hear an old gramophone record playing "Good Night Sweetheart." It had been piped into the loudspeakers instead of the Moscow radio to bid us farewell.

Since my Russian room-mates had departed I had been alone in the compartment and there were few passengers on the car. It was now almost dark, but I had turned off the lights in order to see out better. One of the porters had joined me to watch the crossing.

As we left the glare of the floodlights and moved over the moors I said to him: *"Kitai* (China)?"

"Ne Kitaiskii, ne Russkii," he replied. We were passing a no-man's-land.

The train continued creeping forward for what seemed an age, for I was impatient to reach China. But we had gone probably only a few hundred yards when the porter pointed to a Chinese sentry in a sheepskin coat, carrying a rifle and standing outside a dimly lit guardhouse.

"Kitaiskii," he said, and the train gathered speed. More blackness, then two miles further on we entered a large railway yard and a big, imposing station. It was Manchuli on the Chinese Eastern Railway.

The change was complete. Three miles back every person had been Western and every sign had been in Russian. Now everyone was Oriental and there was not a sign in any language but Chinese. We had entered the Orient as one jumps off a dock into the sea. The only familiar object was one of two large portraits on the station platform. It was Khrushchev. I was to become familiar with the other. It was Mao Tse-tung.

When we reached the large, brilliantly-lit station we were told to wait in our compartments. Our first visitor was a charming and efficient young lady without the slightest knowledge of English. She eventually conveyed to me that she wanted to see my inoculation certificate, but how she understood it I don't know. She was succeeded by polite passport and customs officers. I had a long chat with each of them, if you can call two mutually unintelligible conversations by that name. I filled out a trilingual form for the customs officer, and we then had a discussion over films, cameras and luggage.

He was fascinated by the method of enclosing my colour film in cassettes and putting these into tin boxes and then into bags. He also started to search my baggage and got as far as the two top items — a copy in French of the picture magazine *L'Union Soviétique,* which he looked at from cover to cover, and a bundle of Moscow postcards and colour photographs of my family. These delighted him and he evidently concluded that although my baggage might be irregular it was all harmless.

I had to go to the station and sign a form about the films

I carried, and I was allowed to cash the traveller's cheques the Russians had sold me two hours before.

All the Chinese from the train were inside the station and *so* happy to be home again. I wanted to send postcards to my family and they could not have been more charming and helpful. The post office was shut but they routed out an old woman who sold me some stamps. A curious disadvantage of Chinese stamps is the lack of any mucilage on the back, so a further search was necessary to procure a bottle of glue. We got the cards posted just in time to run for the train, my Chinese friends happy that a stranger's first problem in their country had been satisfactorily solved.

For such an apparently peaceful frontier both sides seemed to be uncommonly cautious.

I got on board to be most civilly greeted by a man in a blue cap and uniform and a red arm band. At first I thought he was a security policeman but it turned out he was only an additional Chinese sleeping-car porter. The Russian porters were still with us. In very broken English the new porter suggested that I might be hungry and that there was a good Chinese restaurant car with European food. I understood his point at once. If I locked myself in my darkened compartment I might see the bare and empty hills along the border, but in the lighted and curtained restaurant this would be impossible. I was hungry and had no desire to watch the dark moors.

I got him to order me a Chinese dinner. It was delicious. Thereafter, until I left China, I ate nothing but Chinese food, using chopsticks.

When I returned to my compartment in the relatively empty train, it was a relief to be alone. I did not miss the air-force officer or his son; in particular I was relieved that his wife had left. Any advantage there might have been in waking up with an uninterrupted view of the attractive wife of a Russian officer in bed four feet away was offset by the recollection that the Russian officer was in bed four feet above and the realization that he was doubtless awake.

But my peace did not last long. In the middle of the night the porter wakened me to indicate that three other passengers had boarded the train and would be in this compartment. Behind him crowded in three Chinese men, each with a suitcase, a bag and a mug with a toothbrush in it. One of them carried a large potted fuchsia.

Clearly they had never been in a sleeper before and they were very excited. While the porter went for bedding, one of them swarmed into the upper berth opposite to me and from there stowed away baggage in the space above the corridor. The fuchsia plant sat on the table.

Bedding in Russian trains comes in two packages per berth. One roll has a pillow, a blanket, and a thin mattress to unroll on the seat. The other contains linen.

While the porter struggled in the crowded compartment to make up two of the berths, I, lying in my lower, had a splendid view of the activity in the opposite upper. Can you imagine an enthusiastic, excited Chinese who has never before seen European bedding trying to make up an upper berth in which he is squatting in a swaying train? He did not know what the towel was for; he did not at first recognize that the pillow slip was a bag; he had no idea of where the sheets went; but he was a quick learner and by leaning perilously over the edge he could catch what the porter was doing to the berth below. Eventually he made it. He got the blanket, himself and all his clothes between two sheets.

For the first time in my life I almost cried myself to sleep —with laughter—before they switched off the light and the Moscow radio.

FIVE

Manchurian Baedeker

On Monday, August 25th, I awakened to my first full day in China. The train was passing through the last hills of the border country. It was still cold and mist shrouded the cliffs along the narrow river valley. We cleared the last gorges and ran down broadening valleys to the swamps and plains of northern Manchuria.

The north of China is not so thickly populated, nor does it resemble the fertile plains I was to see in the south. The extensive fields were richer than those of Siberia but at first the only crop was hay.

Throughout the day it was prairie country with few trees and these for the most part looked like cottonwood. Although hot by noon in August, it was clearly a land of long bleak winters for there were numerous piles of snow fences by the railway. Strenuous efforts were being made to establish a natural snow barrier by plantations of native bushes and poplar trees for miles beside the track. I had noticed similar efforts in Siberia.

Many areas were still uncultivated and I saw one tractor which I thought was breaking virgin ground. The villages were some distance apart, but where we passed people there were many more of them than in Siberia.

Many men were working along the route laying additional sidings, but the traffic on all Chinese railways seemed light after the great density on the Trans-Siberian Railway. Our train continued to run on time as far as I could tell and it now seemed to be going faster which the less dense traffic allowed it to do. To increase the speed the Chinese locomotives which pulled it had much larger driving wheels.

For a long time we passed vast areas of swamp and shallow lakes where the only inhabitants were fishermen. From small boats and by wading they tended intricate traps constructed, like mazes, of miles of bamboo fence set in the shallow water. Their happy children swam in pools, slid down mud banks and scampered along the foot paths over the low hills where stood the tattered settlements.

I noticed two cemeteries, one of which had crosses in it.

In the afternoon we emerged into richer land and passed through several large cities. I could see many factories, buildings, chimneys and more being built, but from the rapidly moving train one had little opportunity to study them. Perhaps I was unobservant.

The fields grew richer and more fertile and along the railway line, especially at bridges, there were ruined pill-boxes left from the Sino-Japanese war.

On this side of the border it was more difficult for me to tell where we were, for there was never a sign in any language except Chinese and these were completely unintelligible to me. My railway time-table and that posted in the corridor listed no stations in China except Peking. I could, however, make some sense out of the two Russian porters who stayed on our sleeper all the way. At the larger stations I walked up and down the platform. If they indicated that there was time I even ventured through the tunnels under the tracks to visit the main stations and other platforms.

These activities, and my enthusiasm for taking photographs, were clearly frowned on by the police who patrolled every platform. Frequently they followed me around to discourage my idiosyncrasies.

All the main stations were like those in rural England, with the tracks open to the air and canopies over the broad platforms. As we penetrated further into China these were thronged with increasing hordes of Chinese. All of them must have had some business in the stations for there was strict control at the entrances. Outside, even larger throngs leaned against the fences to watch the trains go by.

While we waited, at one station (I think Harbin), another train drew in and disgorged an incredible number of young people who appeared to be immigrants or workers moving north. Each had a string bag or a bundle which very probably held all his bedding and personal belongings—a quilt to sleep in and a change or two of cotton clothes. At the bottom of most string bags there was a tin washpan.

This throng tumbled out of the train in which they had been densely packed and hurried up the platform in high excitement to be met by a delegation in the station.

The square outside the station was thronged with people carrying red flags and banners, singing songs and banging gongs in an impromptu band. Clearly this was an organized migration, not merely the arrival of an ordinary passenger train.

I had noticed the band before the migrants' train arrived. Indeed I had managed to photograph it before two white-coated policemen noticed me on the wrong platform and marched in my direction. I thought it wise to retreat to my own train.

The policemen were dressed in dark trousers, white cotton coats and peaked military caps with white covers and green bands around them. They were always armed with very large revolvers slung at their belts. I never saw them behave brutally, although they were brisk and officious. As with our own police, their mere presence seemed to have the required effect.

Later, on the street-corners of Peking and other cities, I frequently saw men of the same force directing traffic. Only

41

on one occasion did I see anyone arrested and he seemed to be an unfortunate beggar trying to ride a train without a ticket.

None of them ever molested or even spoke to me. Although I was anxious to do and see whatever I could, I tried to avoid anything foolhardy or obviously forbidden. If I wanted to take pictures I did so quickly, but quite openly and never of subjects to which I thought they might take exception.

The police knew that my papers must be in order or I could not have got on the train. Certainly in Manchuria, and probably elsewhere, they assumed I was Russian like the other Caucasian passengers.

As the train rattled rapidly through the countryside, I watched the railway workers. The line was an old one and, except for the rolling stock, nothing was mechanized. Men with poles carried heavy machinery. Other gangs shovelled coal by stages onto high platforms and then scooped it into tenders while the train waited. They had to work very hard and the sweat glistened on their bare backs and muscles. The weather was noticeably warmer.

I was interested to notice that women were not employed at heavy work on the railway. I was told that they have been much emancipated in China and can claim jobs which would have been unthinkable ten years ago. Many of the guards, ticket collectors and conductors on trains were girls, but I did not see a single woman working in a track gang.

This is quite different from Siberia, where plenty of strapping lasses did the lighter kinds of manual work such as clambering on top of the sleepers to fill the tanks with fresh water. Of course, the average Russian girl is about twice the weight of the Chinese and I noticed that at six feet one inch, I stood head and shoulders above the Chinese crowds. In Russia and Siberia, as in Canada, I was regarded as being of normal stature.

At Changchun my three Chinese friends departed and

were replaced by two very pleasant German automobile salesmen. One of them spoke excellent English and had a bottle of Scotch whiskey. He was doubly welcomed. It was a relief to be able to speak freely after a week of sign language and monosyllables. We chiefly discussed two topics: the German automobile industry and how to stay healthy in China. In spite of his pleasant disposition, the German's attitude on both these subjects was pessimistic.

"Although the Chinese are buying some automobiles," he said, "they do not buy many and the future market is not promising because they will soon be making their own."

He told me that he had just visited a factory in Manchuria where they made trucks.

"It was an old Russian factory," he said, "which the Russians dismantled and sold to China, probably for a very high price. They have started to make standard Russian trucks and the Chinese are getting on with this job pretty well. I saw no Russians in the plant, but I suspect that there are still some Russian technical advisers about.

"The Chinese are very proud of the fact that they have started to make passenger cars, but I think that these are virtually handmade. The output must be extremely small."

I think he was right, because in Peking I made my Chinese colleagues aware of my interest and they were never able to show me one of these cars, although they often looked for one. Twice they thought that they saw one, but on both occasions it turned out to be some foreign make. However, my German companion was sure the Chinese would make rapid progress in industrialization, and he gloomily predicted that the market for German cars in China and in other countries would diminish.

As he discussed life in China, he asked: "Where have you been?"

He was amazed when I answered: "I've been sitting here for a week because I got on in Moscow."

"It would be very interesting to be able to do that," he

43

said. "I flew around by India to Hong Kong. Have you ever been in China before?"

"No," I said, "but I have been around the world through Africa and India and Australia, so I know the East a little."

"Do you have any pills to take when you get dysentery? Certainly you should do," he said. "I will give you some pills. You are sure to need them."

"Sulphaguanidine?" I asked, for I remembered taking them in North Africa during the war.

"Yes," he said, "you must have some. I would recommend that you drink plenty of Chinese tea; never touch water even to brush your teeth; don't eat salads or fresh fruit; but the cooked Chinese food tastes delicious and is nearly always safe. The Chinese serve it piping hot and even if there are flies around they dare not settle on it."

He seemed surprised when I told him that on crossing the border I had already started a diet of strictly Chinese food, eaten with chopsticks.

He asked me what I had eaten.

"It was really rather a toss-up because I don't understand Chinese at all nor the Russian names for Chinese dishes," I explained. "Last night I had some kind of chicken with rice, rice wine and a bottle of beer. For breakfast I was lucky enough to get an omelette and tea, and being thirsty and not trusting the water, I had a bottle of beer as well. For lunch I had been rather surprised to find borshch put in front of me. The waiter then brought an excellent salad and the complete carcass of a very decayed fish with an enormous mouth."

"You should be much more careful," said my gloomy friend. "Those fish are from the Yangtze River. So far north as this they are liable to be bad, and, of course, the salad was very dangerous."

I said I would at once reform and I gladly helped him and his companion finish the bottle of Scotch, which we all regarded as quite safe. In fact, throughout the East, I wel-

44

comed the prevailing myth that it has great medicinal value. Unfortunately, today it is hard to get and rice brandy is scarcely as delicious, although it may be equally effective as a therapeutic.

When we had finished the whisky, the German got out his Bible, read a few chapters and turned in.

Entry to Peking

Tuesday, August 26th

The next morning I awoke as we approached Tientsin, the old port of Peking. The train was running beside the Yellow Sea, and the reflection of the rising, golden sun mirrored and shimmering upon its oily surface made its name indeed appropriate.

The ocean was calm, dazzling, beautiful. On the other side were swamps and marshes into which pushed fingers of farm land and shallow pans in which the Chinese were evaporating salt.

A canal paralleled the track and along it laden barges crept slowly forward. The land was much richer, indeed very fertile, and every spare foot was occupied.

On board the train our car was alive, bustling with expectation. The porters brought us tea, which we had for breakfast with some biscuits provided by the kindly Germans. They took the sheets and bedding away and brought the baggage down from the racks and out from under the berths.

The Russians discarded their day coats and pyjamas and put on uniforms and business suits and dresses, so that they looked like the technicians and officials which they were.

The Chinese students were tremendously excited. Everyone was moving up and down the corridors except the two Germans and myself. We sat sipping black Russian tea in glasses and nibbling biscuits, while through the windows I beheld the splendid pattern of an ancient land unfold.

After we passed Tientsin I found myself watching for the silhouettes of temples, castles or pagodas of Peking, but the Germans told me that it was still two or three hours away.

I thanked the two blond Russian boys who had been our porters. They had really looked after us well. During the eight days I had come to know them well and could at last distinguish them. They looked so much alike that for the first two or three days I had not realized that there was more than one on the car, for they relieved one another in shifts so that each could get some sleep.

Rather reluctantly one of them accepted a tip from me. I am sure it was forbidden but human nature is stronger than regulations.

Then with a rush we entered the city, flashing past walls of factories, and almost unexpectedly the Moscow Express ended its long journey in the Peking Station at 11.45 — exactly on time.

While I was still collecting my belongings, a neatly dressed and efficient Chinese pushed his way into the compartment and in perfect English inquired:

"Dr. Wilson I presume?"

I bowed.

"How do you do?" he said. "My name is Tien Yu-san, your interpreter. Welcome to Peking. I trust that you have had a good journey."

He took my bags and led me through the confusion on the platform to my hosts, a group of Chinese scientists and officials of the Academy of Sciences of China, who stood waiting to greet me.

I had met none of them before, nor anyone else in China, but they were friendly and welcomed me warmly. All of

47

them spoke some English, and I soon found that they were Dr. Lee Shan-pang, a seismologist, and Dr. Chow Kai, a geodesist, together with a lady and gentleman, he representing the staff of the Academy and she the Chinese Intourist. They handed me three letters from my wife.

By the time the formalities and introductions had been completed, the crowds had filtered away and the hubbub on the platform had subsided. Chatting pleasantly, we made our leisurely way around the turmoil of the station to a side door where we emerged into the sunshine of a Peking morning and a touch of Chinese fantasy. Tall hibiscus plants in pots stood like an honour guard in rows down the steps leading to the calm of an almost empty parking lot. Beyond the gates the crowds streamed past on foot; bicycles and pedicabs, buses and trucks, horses and carts constituted the rest of the Peking traffic.

Through these crowded streets our party drove, in two Russian Pobeda cars, the few blocks to the Hotel Peking.

Throughout my visit I found the Chinese whom I met to be charming, gracious people. My welcoming committee knew that after eight days on the train, there was nothing that I wanted more than a bath, so they escorted me immediately to Room 528 in a new wing of the hotel and suggested that I join them downstairs in three quarters of an hour for lunch.

The Peking Hotel was a vast place. It was modern in the Western style, seven stories high with excellent elevators and many restaurants, a roof garden, lots of shops selling postage stamps and curios, barber shops, telegraph offices, an Intourist desk and every type of facility one could expect.

I discovered from a booklet in the desk drawer that the oldest part had been built in 1915 and other large wings had been added in 1929 and 1952.

My room was designed like many in North American hotels, with an entrance corridor off which opened a tiled bathroom. There were twin beds, and the window looked out over the city. There was a cupboard in the wall for coats, but, in European style, there was also a large wooden chest of drawers. The room had a desk, two or three small tables, two easy chairs, a telephone and electric lights. On the desk was a calendar all in Chinese except for Arabic numbers. This they turned every day.

In the bathroom the water was hot; there were towels and plenty of soap. With the greatest pleasure I had a bath and then sorted out my dishevelled clothes.

Downstairs I found Dr. Lee, Dr. Chow and Mr. Tien waiting in the lobby. All three spoke excellent English and immediately I felt at ease in their company.

Dr. Lee is an internationally known seismologist whose latest books of a year or two ago, although published in Chinese, have been favourably reviewed in the Bulletin of the Seismological Society of America. He is a charming and self-effacing man with a keen sense of humour and a great capacity for enjoying life.

Dr. Chow Kai is a geodesist with whom I had corresponded. He has travelled widely in the United States. Although I found him occasionally a bit sensitive and quick to infer criticism where I had not intended any, I got on with him very well.

Both these men were very able. Like all middle-aged Chinese, they were educated before the Communists took over but I never had occasion to discuss with them their attitude to the new regime. Neither did they ever ask me whether I approved of imperialism and free enterprise.

I knew from corresponding with them that they were scientists. In fact, I knew Dr. Lee's work quite well but I was soon able to establish that they were indeed working scientists in another and rather unexpected way.

For lunch they took me first to a dining-room on the top floor, but when I saw that the few people in it were mostly European and that all the places were laid with knives and forks, I realized that here Western food was served.

I explained to my friends:

"While I am in China I have resolved to eat Chinese dishes with chopsticks whenever possible. Can we go to a Chinese restaurant?"

They agreed with enthusiasm but were thrown into considerable confusion. It was perfectly clear that none of them had the least idea where to go to find the other restaurant. Mr. Tien had to ask several people before we found the dining room in which Chinese food was served to foreign visitors. It was on the ground floor at the other end of the building.

Had these men been professional guides assigned to "watch" a visiting foreigner, they could hardly have failed to know their way about the largest hotel in Peking. Thus I was assured I could accept them at face value.

I later discovered that there were many dining-rooms in the hotel including one for Moslems and others for Chinese, but the one in which we found ourselves was small and most interesting. It was peopled by an astonishing variety of races—Indians, Indonesians, Brazilians, Australians, English, French and Canadians, many of them with their Chinese hosts.

I went there regularly thereafter. I found the food delicious. In this small room I could hardly avoid hearing some of the other conversations and I found this entertaining when I was eating alone for I could understand most of the talk. (In what language but English do you suppose that the Chinese and most of their foreign guests could converse?)

On this first day we had an excellent lunch and became merry over beer and rice brandy. Mr. Tien introduced a "Toast to Peace," which reminded me of the importance attached by Communists to this gesture. I amused myself for the rest of the trip by trying to offer this toast before they did. It seemed a harmless piece of one-upmanship to suggest that the Western countries were at least as desirous of peace as the Eastern.

After lunch we went to my room for an informal chat about my trip in China. They suggested that I spend ten days in Peking seeing universities, geophysical institutes, and surveys and visiting a dam, a collective farm, the Great Wall and old palaces.

They felt that I might then spend two or three days each at Shanghai, Nanking, Hanchow and Canton before leaving by Hong Kong.

I indicated my pleasure at their suggestions, but I was non-committal. People who had known China—especially Professor William Dobson, Head of the Department of East Asiatic Studies at the University of Toronto—had urged me to endeavour to go up the Yellow River, if I could possibly do so. Dobson wanted to know if the famous relics of Confucius there were still safe. The country sounded in every way more interesting than that along the coast. However, I withheld any mention of this plan until I was more confident of my status in the country.

I asked Mr. Tien if I could register at the British Embassy and he took me there by car although it was only two or three blocks from the hotel.

The British have occupied this Embassy for nearly a century. It is a fascinating and splendid place. It had belonged to a leading member of the aristocracy who became bankrupt about 1860, and sold the British his ducal palace.

It occupies an entire city block and is surrounded by a great twelve-foot wall. Within are a number of red-pillared pavilions in the Ming palace style surrounded by glades and gardens. Scattered among these are the modern embassy buildings accommodating the chargé d'affaires and members of his staff.

At the entrance on this first visit I noticed the outside walls were plastered from top to bottom with thousands of streamers of coloured paper overlapping one another so that

51

the wall looked like a dirty old feathered cloak. I did not understand the significance of this nor had I the time to speculate upon it before the Embassy doorman, an old Chinese who was being watched by a young policeman, admitted me and directed me to the offices.

I walked there past the lovely scarlet columns of the buildings and through trees and parks where English children played attended by their Chinese nurses.

At the office a retired sergeant-major, still very regimental, checked my papers, found me five more letters and handed me on to one of the secretaries. The latter explained that his wife and most of the staff had gone to the seaside for their holidays, but eventually another secretary was found whose wife was at home. She provided us with delicious tea which I doubly enjoyed for the pleasant relaxation after eight days and nights on the Trans-Siberian Railway.

I should mention here how grateful I am to the Canadian and British Embassies in Moscow and Peking and elsewhere on my travels for those few opportunities to relax. Although I had no very arduous duties to perform, I must admit that much travelling gave one a slight feeling of tension.

The Russians and Chinese whom I met were most friendly, but I suppose that in the back of one's mind was constantly the thought that between their governments and my own an unfortunate state of passive hostility existed.

While I sipped tea with my compatriots I learned that, only a few weeks before, the Embassy had been subjected to a thirty-six-hour massive verbal barrage. During this demonstration, which the Chinese Government claimed to have been spontaneous, the walls had been plastered with defamatory posters, and it was the tattered remnants of these I had noticed as I arrived.

Leaving the Embassy I looked more closely. Most posters were in Chinese but a few in English said: "Go Home

British Imperialists," and similar and more offensive remarks.

The streets around the Embassy had been completely blocked with people who kept up a continuous abusive chanting. Many of them banged gongs while loudspeakers blared. The demonstrators had brought tens of thousands of posters with which they covered the walls from the ground twelve feet up to the top and all the way around the block. They were all colours — yellow, red, purple and white paper on which the messages were scrawled in large black Chinese ideographs.

After a few weeks and a rainstorm or two the wall appeared most bedraggled. It made a bizarre scene outside any Embassy, but the British were too astute to take the posters down and the authorities clearly did not want to. Like a highwayman on a gibbet, the posters seemed destined to hang forlorn until they rotted and fell away.

Poor Mr. Tien waited dutifully outside. I felt rather guilty when I came out, but he bore me no resentment and took me for a drive through some of the principal streets of the old walled city. In one place a new road was being built through an area of homes. I was told that some streets had been widened and many had been paved.

A few buildings within the old city had been demolished and replaced, but most of the new buildings were outside the walls. These magnificent and ancient mud walls, already breached by a number of boulevards, were fast being demolished. It was only a matter of time until they would vanish entirely.

Beyond the walls, in what had been vegetable gardens and fields, a host of new offices, institutes, factories and ministries were rising. Each was surrounded by its dormitories. I noticed that the Chinese never spoke of their homes, flats, apartments or houses, but always of their dormitories. Every new working place seemed automatically to have beside it enough dormitories to house its workers.

53

Some of the ministries, and the nearly completed television building with a transmission tower on top, were eight or ten stories high and so were the main hotels, which had elevators. Most other buildings were three or four, or occasionally five, stories high and had only staircases.

Some of the ministries had bell-cast roofs covered with lovely tiles and mural decorations on the brickwork, but this was unusual. Most of the buildings were rectangular blocks of grey brick with simple tile roofs. They were well built but strictly utilitarian.

As we drove through the streets I noticed a motley assortment of vehicles: cars, but not enough to constitute a traffic problem, from every country in the world; heavy trucks, mostly Russian, but some European and American; lots of horse-drawn carts that characteristically had two big wheels, unlike the carts in Russia which had four wheels. In addition to a horse between the shafts, a donkey pulling on two ropes commonly goes ahead of the horse.

There were lots of pedicabs and peditrucks — the vehicles built like bicycles and propelled by manpower. There were even pedibuses for school children. All of there were old, dilapidated and not as numerous as in other eastern cities outside of China.

Mr. Tien told me that it was not dignified for humans to act as beasts of burden and that they were making no more pedicabs. He pointed to the new electric and diesel buses which were operating on the main street and which had been added to the old street-car system. He said that as soon as the bus services were complete there would no longer be any need for pedicabs. The pedicabs take two people and are pulled by a built-in bicycle, instead of the running man who pulled the rickshaws, their immediate ancestors. Besides people, on several occasions I saw pedicabs carrying a curious load — two metal cylinders of acetylene and oxygen for welding. It seemed to be the recognized method of transporting those gases.

There were peditrucks everywhere. They were old, and not infrequently they broke down. Often I saw their unfortunate and ragged drivers sitting on the roadside fixing their dilapidated bicycles. There seemed to be no doubt that they were being replaced by trucks, although many were still available to carry an immense variety of loads. Light but bulky goods such as furniture, piled perilously high, were common.

The pedibuses were amusing. I saw them only two or three times, but behind the man who peddled was a miniature bus each holding eight or ten small children. I suppose they were being taken to or from nursery school.

The street-cars were rickety and probably familiar to many old China hands. I noticed that the original cars, numbered 1, 2 and 3, were still operating as were, no doubt, all the other cars put into service since the system was founded. (Incidentally, the use of Western numerals now seems to be universal in China.)

The motor buses were new and good and I noticed that some were Czech Skodas. Once, Mr. Tien positively shouted with pride: "There is a Chinese bus."

"You mean one made in China?" I asked.

"Of course, they are made in Shanghai and they run very well. Soon we will make all our own trucks and cars, and what will be the use of the American imperialists' embargo then?"

What he said was apparently true. In Peking and other cities I saw dozens of these buses, all new and identical and having a name plate in Chinese characters. Later, in the Export Trust Building in Canton, I examined one. It appeared to be a good bus. I have no proof that it was made in China, but it is logical that anyone in the position of the Chinese, founding an automobile industry, would start by building buses and trucks. I do not know, but to say that they import some of the parts is beside the point. If they only make part of the bus today, tomorrow or in ten years they will make it all.

55

Along the route of our tour we stopped once or twice to take pictures and I was immediately surrounded by a crowd of small boys. I became used to this, as Caucasians are so few in China today that they attract attention anywhere, even in the heart of Peking. As Mr. Tien said, "They probably think you are a Russian."

On the drive we passed one of Russia's gifts to China — a gateway whose design was typically Russian, wedding-cake style like Moscow's skyscrapers and as ornate as a 1959 American car. In front of it was another touch reminiscent of the Soviet Union — several large plaster statues of athletic young men and buxom girls dancing in postures of gaiety and enthusiasm. Although the figures surpassed nature, I felt somehow that they would not have commended themselves to Western critics of art.

It was not a subject on which Mr. Tien would express his views. He said only: "Inside the gate is a permanent exhibit of Russian goods".

However, I noticed that he made no suggestion that we should visit it nor did he ever show me any other Russian things.

He contented himself with remarking: "The exhibits demonstrate for us the progress possible in a vital Communist state, where everything is devoted to the welfare of the people. It illustrates the love and unity existing between Russian and Chinese people. It is a pity that all men cannot be brothers."

I replied as solemnly as I could: "That is a very interesting and splendid thought, Mr. Tien, and you have expressed my own feelings beautifully, but since I am after all just a decadent and pleasure-loving Westerner, you will therefore not be surprised if I enquire whether we can go shopping to buy something Chinese for my wife and daughters. Later we may be too busy".

Before coming to Peking I had been told that it was a wonderful place to shop. Moscow had not been, because for forty years the Russian people have had more on their minds than making trinkets for tourists. The Chinese still had a lot of things left from more opulent times.

Mr. Tien took me first to the new department store.

He said: "Strange as it might seem to you, there was no department store in Peking until the new regime built one. You can see how backward the country was and the great progress which has been made in only eight years since the glorious liberation."

I quickly became accustomed to these sermons. I came to expect them and sometimes I would beat him to it. Recognizing something new, I would suggest the great achievements of the new regime to him. In this amiable fashion the conscientious Mr. Tien and I rambled through the ancient and beautiful kingdom of China.

He probably regarded me as flippant but at least it must have encouraged him to feel that I could see and recognize great achievements when I saw them, even if I did not treat them with the seriousness and awe he thought they deserved.

The department-store building was three or four stories tall and appeared to be less than eight years old. Inside it was completely Western but conservative. On the top floor it had a particularly lovely selection of new lacquer work, *cloisonné*, carved wood, woven bamboo, porcelain, bronzes, printed silk and carved ornamental stones. In fact every kind of pleasant Oriental knick-knack was available. Both the service and the goods seemed to be excellent. Many identical articles can be found for sale in Canada but at ten times the cost.

I bought some little carved goldfish of rosewood in Peking for 17 cents each. After my return my sharp-eyed fourteen-year-old daughter Pat discovered identical goldfish on sale in Toronto for $1.65.

The counters along one whole wall of this floor were devoted to the sale of lengths of silk. I looked at a number of rolls, but with the exception of one wash print, which I later bought, I found them a bit gaudy. The wash print I bought bore the modern design in which the pattern is composed of intricate scribbling — many continuous lines each of a different colour.

On the second floor were ready-made clothes and on a later occasion I bought two silk shirts there, but I chose them too quickly and they were not wholly successful. The fault was mine.

Downstairs, they had, among other things, a lot of simple furniture, pottery, thermos bottles and household goods. It was apparent that most goods in the shop, if not all, were made in China.

There were a few Chinese in the shop, but it would not be considered crowded for a department store in any of our cities of comparable size. So far as I could observe I was the only foreigner among the shoppers. Foreigners of any nation including Russian are rare except in the hotels.

I made a list of the prices of some articles which, at the realistic and legal rate of $2\frac{1}{2}$ yuan to the dollar, were as follows:

Turquoise necklaces	about $25.00
Jade necklaces	$50.00 to $280.00
Large amethyst brooch	$16.00
Jade rings	$10.00 to $75.00
Taffeta silk and silk prints	$2.00 per yard
Blue enamel dishes (15″)	$12.00
Enamel plates	$4.00
Enamel vases	$4.00
Red lacquer cigar boxes	$6.00 to $8.00

Next we went across the street and a block away to the old market. This was much more picturesque. A series of sheds provided cover for a vast bazaar. Innumerable stalls each with its own shopkeeper dealt with a wide variety of commodities. I suppose they all belonged to some Communist trust but the effect was still that of a great group of independent traders. This was undoubtedly how the bazaar had begun.

They sold a most astonishing variety of goods — clothes, paper umbrellas, shoes, pottery, jewellery, stationery, food, fountain pens, electrical appliances, and household furniture and furnishings. The prices were all fixed. Everything had a price tag on it and there was no haggling. I'm not an experienced shopper but everything seemed to be extremely cheap.

I was looking specially for things I might buy as presents. Again I found the silk disappointing. There was lots of jade, lovely turquoise and many kinds of carved stones. Some of these seemed very cheap, and on a return visit I bought several old pendants and buckles in carved moss agate and in rose quartz for prices varying from 25 cents to $1.00.

I looked at the jade with interest but I knew so little about it that I felt it was inadvisable to get any. They also had star sapphires and other expensive gems but it seemed to me that in this market they had only second quality and that the finest gems would be found in more exclusive shops.

The bazaar was thronged with Chinese and all the merchants were polite and attentive. The jewellers in particular looked on me as a hopeful customer and took me into their inner rooms to show me their treasures.

In both shops I furiously noted down prices, but at that time bought nothing. Faced with such a wealth of lovely things at such cheap prices and having only limited funds and still more limited room in my luggage I wanted time to decide.

When we went outside it was getting dark. The lights were coming on as thousands of gentle, blue-clad Chinese were hurrying through the streets. There were strange shouts, strange scents, oriental music and no one in sight but Chinese. I found it all exciting and delightful, but I was not sorry to return to the hotel for a late supper and a long sleep.

59

SEVEN

Plans, palaces and Peking duck

Wednesday, August 27th

I did have an excellent sleep, and thus fortified I went downstairs to tackle breakfast in the Chinese restaurant. I always found this meal the most formidable of the day, and sometimes, on the pretext of joining someone else, I ate it in the European restaurant; but this morning I resolutely set upon the stewed rice, sardines, sour pickles, tea, and steamed bread that constituted a standard Chinese breakfast.

I was fascinated to discover that half the Chinese people do not eat much rice, which does not grow in the Yellow River Valley or in the northern part of the country. They are wheat eaters, but their bread is quite unlike our own. The leavened dough is steamed and served as damp, grey-skinned dumplings. I did not find these particularly palatable, so I ate rice; but Mr. Tien, who came from Honan, preferred them.

At ten o'clock a delegation was due to arrive to discuss formally with me the program for my trip in China, which had been touched upon lightly at lunch the day before.

My conversations in the British Embassy had reinforced

my desire to see the Yellow River Valley. The idea of travelling along the old Silk Road to Europe fascinated me. I might see the mountains of Tibet, the turbulent Yellow River flooding through the terraces of golden loess blown through the ages from the Gobi desert to coat the hills and valleys with a hundred feet of caked yellow dust. It was surely an exciting land, stirred alike by great earthquakes and by enterprise both old and new. There lay Sian, the ancient capital of the Chin emperors who had first united and named China 2,300 years ago. There lay Lanchow, the ancient gateway and the new strategic centre of the country. I wondered how I could substitute visits to these places for the more conventional trip through the coastal cities that my hosts had proposed. I hoped I would be successful, because all my interests attracted me to the interior.

I came downstairs at ten o'clock and was greeted by Mr. Tien, but it was not until ten or fifteen minutes later that Dr. Lee and Dr. Chow arrived, bolstered by the added authority of their companion Dr. Chen Tsung-chi, the Vice-President of the Chinese Committee on Geophysics. I surmised that this slight delay was intended as an indication of independence on the part of the Chinese.

The program which they politely and gradually unfolded during the course of the next hour was quite detailed. This is typical, of course, of all Communist countries. There must always be a plan and everyone must follow the plan, whether it is a five-year plan for remaking a country, or arrangements for a picnic.

The program they proposed was as follows:

WEDNESDAY, AUGUST 27TH: After this morning's meeting, a visit in the afternoon to the Forbidden City, and a Peking duck dinner that the Secretary-General of the Academia Sinica proposed giving in my honour.

THURSDAY, AUGUST 28TH: Inspect the Institute of Geophysics and Meteorology, prepare and write out two speeches, view the Temple of Heaven, attend a Chinese opera.

61

FRIDAY, AUGUST 29TH: Deliver the first lecture and drive to the Central Geophysical Observatory outside Peking.

SATURDAY, AUGUST 30TH: Tour the Great Wall, the tombs of the Ming Emperors and a dam.

SUNDAY, AUGUST 31ST: Motor out to Choukoutien to the dig at the caves of Peking Man.

MONDAY, SEPTEMBER 1ST: See the Institute of Geology and deliver a second lecture.

TUESDAY, SEPTEMBER 2ND: Visit some departments of the University of Peking and see the Summer Palace.

WEDNESDAY, SEPTEMBER 3RD: Go through the Institute of Geophysical Prospecting and catch the late-afternoon train to Nanking.

THURSDAY, SEPTEMBER 4TH: Arrive in Nanking and visit the astronomical observatory.

FRIDAY, SEPTEMBER 5TH: Sight-seeing in Nanking.

SATURDAY, SEPTEMBER 6TH: On the train from Nanking to Shanghai.

SUNDAY, SEPTEMBER 7TH: Go through the Zikawei astronomical observatory at Shanghai.

MONDAY, SEPTEMBER 8TH: Leave Shanghai by train for Hanchow.

TUESDAY, SEPTEMBER 9TH: Arrive at Hanchow and go sight-seeing at the Western Lake.

WEDNESDAY, SEPTEMBER 10TH: Take the train to Canton.

THURSDAY, SEPTEMBER 11TH: On the train.

FRIDAY, SEPTEMBER 12TH: In Canton.

SATURDAY, SEPTEMBER 13TH: Leave for Hong Kong.

I listened to this program with the greatest interest and every mark of respect. I wrote it all down solemnly, with many expressions of my genuine interest and delight.

When they had concluded I said: "No program you could have planned for Peking would please me more, because I have come to see geophysics and it is clear that you intend to show me the most important geophysical institutes and observatories and university departments.

"With regard to the second part of the trip, I am sure that what you have proposed would also be exceptionally fascinating; and if pleasure and interest were the sole reason that brought me to China I cannot imagine anything more delightful than to visit these famous cities and such beauty spots as the well-known Western Lake at Hanchow."

Then, chancing my arm: "However, ignorant as I am of Chinese geography, it seems to me that the trip you propose lies mostly along the fertile plains of the coast where there is little of geophysical interest. This would be fascinating for anyone to see, and especially for anyone interested in agriculture or farming; but I suppose there would be few rocks there and I know that that part of the country is relatively free of earthquakes.

"It has occurred to me that because the interior is more mountainous it may perhaps offer more of geological interest, and since it is notorious for its numerous earthquakes there might, therefore, be more geophysical observatories and institutes to be seen in the valley of the Yellow River or elsewhere in the interior than along the coast."

Dr. Chen replied: "If you had accepted our original invitation to spend four weeks in China we had hoped to take you to Chungking in the interior by train and to return by boat through the famous gorges of the Yangtze River."

He politely implied, since I had been so misguided as to spend a week on a rather dull train journey across Siberia, when I could have flown to Peking in a day, that I had missed my opportunity to see these famous gorges. I agreed that I had perhaps been unwise but that my foolishness arose from an ignorance of the true beauty of China and from a long-cherished, if misplaced, desire to travel across Asia by the famous Trans-Siberian Railway.

I suggested that although the gorges on the Yangtze were well known even in Canada for their beauty, I unfortunately had not come, and had not time, to see the splendours of the scenery. I asked if I could make some shorter journey

63

to the west which would enable me to see the scientific progress being made there in my own field, even if I could not have the pleasure of traversing the gorges.

"At Lanchow, surely, there are scientific academies and great universities being established by your new and vigorous régime that you would wish me to see. At Sian in the Yellow River Valley is the fabulous country of yellow loess that I wish to visit; and are not the rightly famous relics of Confucius preserved there, and the monuments of the great Chin emperors who first established the full sovereignty of your illustrious and ancient county 2,300 years ago?

"Forgive me for venturing to suggest any alterations in the magnificent plans you have made for me. I am sure that you will know and arrange what is best. If I overreach myself, it is only because I so much appreciate the chance to see this wonderful country and desire to use my limited time to the best advantage; but would you not rather show me Sian and Lanchow than the westernized cities of the coast? Surely you regard them as relics of a foreign, decadent and now vanished imperialism?"

I am sure that we all enjoyed this exchange, which took place politely, sedately and with the best of good humour over many cups of China tea. I felt sure that all these men knew the West well enough to have concluded arrangements in a brisk Western fashion had they so desired, but they enjoyed leisurely Oriental ceremony, and I did too. I am not sure what Mr. Tien thought, but as everyone spoke English he sat as a mute, if puzzled, observer.

In the end, the Peking program was firmly agreed upon and they said they would look into the possibility of making the changes I had proposed. This might have been a piece of face-saving, but I think it much more probable that the scientists with whom I was dealing had no power to make such decisions. As a matter of fact I was never told what the decision was, nor did I seek to assuage my curiosity by asking for an answer. When the last day came in Peking we simply got on a train, and only then did I discover which route we would take. Meanwhile I profoundly hoped that I could see the old Silk Road and at least the rim of the marches of Tibet.

The idea that the Chinese are impassive and inscrutable people seems entirely wrong to me. The different scientists whom I met each day, although serious and hard-working, seemed to be full of good fun and humour. When we relaxed over a meal or on a trip to see some palace, we all enjoyed ourselves thoroughly. They were polite and considerate and seemed genuinely glad to see me, for my visit provided them with opportunity to display their achievements and to get news of the work of others. The fact of the matter is that they gave me a very good time in China.

I believe the inscrutability that has been attributed to the Chinese arises from two causes. One is their former position of inferiority, which they felt keenly; the other is a more fundamental matter, connected with their philosophy and the nature of their language.

Chinese is a language with an unfamiliar grammar and a comparatively small vocabulary. A great wealth of meaning is conveyed by the skilful use of this vocabulary and by giving a multitude of meanings to a single syllable, differentiating these meanings by inflection and tone. By "tone" is meant the manner in which a syllable is pronounced, whether upon a single pitch or with a steadily rising accent, or a falling accent, or an accent that rises and then falls. There are said to be as many as eight different tones in use in some Chinese dialects.

In other languages these tones are not needed for this purpose, for a greater variety of words is available; and inflection is therefore set free to express emotion, rather than to convey meaning. I suspect that a Chinese cannot display so much emotion in his speaking as we can, for to do so would alter the meaning of his words. He has to speak impassively to be correctly understood. It is also traditional good manners to be calm and not to display emotion.

It seems to me that this makes it doubtful whether one can thoroughly understand Chinese culture, or any other, without first knowing the language. On the other hand, English is satisfactory for discussing science anywhere.

I had a light lunch in the Chinese restaurant and received a truly beautiful card, entirely written in Chinese and illustrated with a bough of apple-blossom. Mr. Tien explained to me that it was an invitation for dinner that evening. Soon afterwards, we got away in a car to the famous Forbidden City, the old palace in which the Emperors of China had lived for centuries.

I believe that Peking formerly crowded up to the palace walls and that the vast square that now lies between the Forbidden City and the main railroad station has been but recently cleared. In the centre of it rises a white column, the monument to the People's Heroes of the Revolution. The square has clearly been modelled on the Red Square in Moscow and like it is designed for ceremonial occasions. It gives a fine view of the Tien An Men, the principal gateway to the Forbidden City. Its crimson walls, built by the Ming Emperors 500 years ago, have been freshly painted; and upon them, beneath roofs of shining yellow tile, is a great inscription in white characters facing the square. No doubt it boldly proclaims the virtues of the new régime.

The palace buildings form a vast area set in the heart of Peking and surrounded by a massive wall and moat. At either end there is a principal gateway running like a tunnel through the main walls and admitting one to a succession of spacious courtyards, separated by palaces and pavilions strung like beads across the city, with names like a chime of bells—Hall of Majestic Peace, Palace of Earthly Tranquillity, Hall of Blending of Earth and Heaven, Gate of Heavenly Purity, Hall of Precious Harmony, Gate of Supreme Harmony.

Each courtyard is a sunken oasis of white stone set among the crimson walls of palaces and shadowed pillars of teakwood pavilions. They are floored with cobblestones and paving and brightened with flower beds, plots of grass, and huge decorative bronzes. Some of these are urns bigger than a bathtub, others are fierce heraldic lions and dragons higher than a man's head. Terraces and balustrades of white marble ring them, and thousands of carved posts, representing a prodigious effort by skilled craftsmen, give a light and gay effect.

The stairways leading down from the palaces are intricately decorated with imperial dragons in bas-relief. At the time of my visit, some workmen were laying wooden treads to protect them, while a few others were restoring and maintaining the fabric of the city.

The palaces—great forts with massive red walls—alternate with light pavilions and gates, supported by teak columns. All are surmounted by bell-shaped roofs covered with corrugated and intricate tiles of shining imperial yellow. So numerous are the roofs and so complex their designs that from any vantage point, like the nearby Coal Hill, the succession of curving yellow sheets look like waves of the sea lighted by the richest moment of a golden sunset.

Within each of the principal palaces and pavilions is a great throne room, dominated by a stage upon which rests a giant throne, cushioned in imperial yellow silk richly embroidered and surrounded by feather fans, lacquer screens, ornate *cloisonné* vases, wooden carvings, and all the ceremonial trappings that men could devise through centuries to lend weight, prestige and beauty to the courts of the Son of Heaven. Each had been devoted to some particular facet of the ceremony and ritual attendant on the former government, so highly developed around the lives of the Emperors.

One room was the court where justice was meted out, another where envoys were received, another where victorious generals and heroes were rewarded; but it was impossible for me to do more than guess which was which.

All were now arranged as museums, through which we—

and hundreds of Chinese people, for some tiny sum—were free to wander, examining the glass cases filled with treasures and regalia. In one throne room was a series of elaborate drawings and paintings illustrating the order of procession and the costumes proper for certain rituals and ceremonies that the Emperor had been accustomed to perform. In another room was a collection of the seals used by the Emperor. Instead of signing papers, he stamped them with a seal of carved ivory or stone. Whereas private individuals in China have seals (or chops, as they are called) about the size of one's fingers, the Emperor had seals, for declaring war and for other important matters, as much as eight inches square and very beautifully carved.

Another room, crowded with Chinese sight-seers, contained a collection of seventy-four mechanical clocks, each of which was from two to four feet high. Evidently they had delighted the fancy of some of the Manchu emperors and empresses. Most had been made in Paris or London during the eighteenth or nineteenth centuries and were most ingenious pieces of clockwork. (I think that the Chinese had made some of the later ones.)

At each hour and quarter-hour chimes sounded, music played, little figures came out of trap-doors and danced, mirrors revolved—in fact the various clocks, in their different ways, displayed every kind of mechanical ingenuity one could imagine. All of them were triumphs of the jeweller's art, heavily gilded, set with semi-precious stones, and enamelled.

When we reached the end of the palace we came to a stream crossed by five parallel bridges of white marble and channelled, by intricate canals, into pools and streams in other courtyards. We followed it to the east side of the palace and went back towards the gateway, and our car, through the Empress's quarters. Here the courts were smaller, more secluded, and arranged for luxurious daily life, in contrast to the principal courtyards, which had been designed for ritual. These dark and faded chambers of departed splendour opened out on courtyards shaded with pine trees and scented with hibiscus and jasmine.

In the private rooms of the empresses were displayed large, richly covered beds, set into alcoves like lower berths but appearing much more uncomfortable. Cupboards and cases were filled with ornate and elaborate embroidery. In other rooms were gold and silver dishes, ranging from large golden bowls of simple shape to curious hook-shaped ornaments. These were about a foot long and had some traditional significance. In another case were scores of cups of pure gold bearing the names of members of the nobility, and elsewhere were many jewelled and golden pagodas as much as three or four feet high.

In several of the rooms were cases full of carved jade, the more translucent pieces on stands before the windows, and the larger on separate pedestals. Some of these huge masses were boulders three feet in diameter. It seems that the emperors valued jade for itself. Most of these boulders have not been greatly altered in shape and frequently they have been set on one end and covered with carvings until they resemble mountains. On the top were clouds and forests, on the sides temples, people and stairways. In a few cases vast masses of jade have been cut into urns, richly and deeply carved.

A feeling of injustice that so much wealth had been lavished on so few overcame my sense of melancholy at the wasted beauty around this dead fairyland. We walked out of a side gate and returned to the car. I saw the plaster peeling from the massive outer walls beside the unfrequented path we followed.

This vast display of imperial magnificence recalled me to my duty as a tourist, and I applied myself to the business of acquiring suitable offerings to bring home to my family and my friends. I had heard that the best shops used to be in a place called Embroidery Street, and at my request Mr. Tien made inquiries and took me there. Upstairs in a small shop we were courteously received by three Chinese who appeared as though they might have once been the proprietors, but who were now clearly employees.

What a wealth of things they had! Rolls of silk, more beautiful than those in the department store, were ranged in cases —patterned taffetas in black or white, gay prints, shot silk, and brocades woven like damask in ice blue, light green, mauve, yellow, or other pastel colours. Loveliest of all, they had a thick ivory silk patterned in a most delicate way as a brocade in five vivid colours—scarlet, gold, bright and dark blue, and emerald green. They had piles, feet high, of mandarin coats, embroidered in exquisite fashion, that had once belonged to nobles. From other coats they had cut the elaborate insignia to make wall hangings, table centres and place mats.

The man-centuries of labour involved in all this exquisite needlework, and the delicacy of it, held me at once fascinated and appalled me with its wasted opulence. Some of the silks were very ancient, and the three Chinese explained their faded and delicate colours by long burial in the ground. Those that were oldest were cheapest, for although they were admittedly the rarest, nothing over a century old might be taken out of the country and there was little market within. They had rugs and hats, cushion covers and scrolls, fans, and all the specialized and useless finery of a vanished era. They also had furs; mink, leopard skins, lynx, and sable. I was much too bewildered to know what to buy, but all three salesmen, glad to see a solitary customer, unfolded the treasure of a museum or of an emperor's palace before me. Shaking out their wares, they filled the room with fragrance of sandalwood and camphor.

If I said that my wife liked blue, they pulled blue coats and silks from cases; if I requested something for the children, they had wedding dresses for little Chinese ladies; if I said furs, they opened teakwood chests and displayed them.

70

In my confused state I decided to get nothing that day, but I just wrote down the following prices:

Superb full-length dark blue embroidered mandarin coats	$64.00
Scarlet, white or black embroidered two-piece wedding dresses	$7.20–$11.20
Chinese mink stoles	$102–$144
Sable stoles (not made up)	
24 skins	$120
30 skins	$156
Brocade in six colours	$2.05 a yard
Silk prints	$1.26 a yard
White or black taffetas	$1.31 a yard
Fuji silk shirts for men, made to measure	$5.60

We returned rather dazed to the Peking Hotel and I took a hot bath to refresh myself for the Peking duck dinner of which Mr. Tien had told me and for which we were due to be picked up at 7 o'clock.

I was ready on time, but Mr. Tien appeared and said that he had just had a telephone call, saying there would be a delay of perhaps as much as half an hour because the restaurant was full. I felt sure that there might be some significance attached to this, but I did not show it. In fifteen minutes or less another call cleared our way.

It took only a few minutes in the car to cross the square and pass along the great east-west boulevard to a Moslem restaurant, famous for centuries for its Peking duck dinners. It was a small place, ancient and not particularly prepossessing. We walked down a corridor, past kitchens, into a well-lit anteroom. There I met Dr. Pei Li-shan, the secretary-general of the Academia Sinica; Dr. Lee, Dr. Chow and Dr. Chen; Dr. Chang Wen-yu, the vice-director of the Geological Institute; also the secretary of the Institute for Relations with Foreign Scientists and a young lady from the same Institute, whose names I did not catch.

The austere and commanding Dr. Pei Li-shan said to me:

71

"Scientists are welcome in China. We are particularly glad to greet you as the first member of the National Research Council of Canada to visit Peking during the past eight years, but we welcome you with special warmth coming as you do from the land of Dr. Bethune."

This reference to Canada as the land of Dr. Bethune flabbergasted me; but I replied that I was duly grateful, and said:

"I bring you greetings from other Canadian scientists, and I am sure that the National Research Council will be glad to hear of this meeting and the honour you have done me. I am sorry that although Dr. Bethune was a graduate of the University of Toronto, he had left before I came, and I never had the pleasure of meeting him. It is gratifying to know that a Canadian is remembered and honoured in China for his medical work."

I must admit that I had only vaguely heard of Dr. Bethune, but I later discovered that after a serious illness in his youth, he had been seized with a vital urge to help mankind. His interest in the unfortunate, more than his political beliefs, had taken him to the civil wars in Spain and in China to work as a surgeon, asking little and giving much, while he worked in conditions of misery, lacking all but the most primitive facilities.

A chance infection had killed him in China, where he is particularly remembered for his success in setting up blood banks and transfusion services. To do this, he had had to overcome the great prejudice of Chinese against giving their blood. He had often given his own as an example. I was to hear much more of him, because his name and his work are held in high regard by the Chinese. I was told that they have erected a memorial and named a hospital in his honour.

After the introduction, a cup of tea, and a cigarette, we entered the adjoining dining-room, and I sat down at a circular table between Dr. Pei Li-shan and Mr. Tien, who leaned over to interpret.

With cold spiced meats and hors-d'oeuvres, we began the

72

feast. It consisted of twelve courses, each made of some part of specially reared ducks; but no one who had not been told this would realize it, for the dishes were well disguised with sauces and pickles, and the different pieces of duck were hidden in soups and in patties and served with a great variety of vegetables and wines.

Unfortunately, I neither recognized nor can I remember the full list of those delicacies, but I do recall that our toasts began in Chinese grape wine (which I found delicious and which would have contented me through the meal), were continued in white rice wine, in rice brandy, in beer, and in hot yellow rice wine!

The most exotic of the dishes was perhaps the soup of ducks' tongues; but the main course, for which I had been warned to save myself, was the roast duck. These delicious birds, roasted to a rich, dark brown, were brought in whole and carved with a vast knife with the utmost speed and skill by a woman cook. Pieces dipped in a rich sauce were dropped into holes made in the sides of special buns, or wrapped in thin, delicate pancakes and eaten with the fingers.

Every part of the meal was completely delicious, and I can well believe the story that the high reputation of French cooking stems from the fact that the Jesuits brought home some of the key receipes from China. Finally, when we had exhausted every gustatory delight that the best cooks could concoct from a duck, we had delicious pineapple and the largest peaches I had ever seen. These, Dr. Pei Li-shan said, had been brought especially from the extreme south of the country.

We all drank toasts to peace, to China, to Canada, to friendly relations among scientists, to geophysics, to one another, to the success of my visit, to my safe return, to our next reunion, to absent friends, to the Chinese Academy and to everything else that anyone present could think of. I thought it would not go amiss and proposed a second toast to peace. We drank these in all the liquors mentioned, but fortunately the glasses were small and the wines not particularly strong.

I found that my hosts had a habit of standing up one after another and toasting me alone, bottoms-up, but being cautious of the effect this might have if I were to drink more than any of the others, I generally made the excuse that each toast was so important that we should all rise and drink taps to it.

The thought that I might get somewhat tipsy did not dismay me in the least. I was only fearful lest I should get drunker than anyone else. As a matter of fact, although one of the smallest Chinese gentlemen took to giggling a good deal, we were all able to walk to the anteroom for tea and cigarettes and the serious discussion of the evening.

I considered that if I had not distinguished myself, I had at least survived. The fact that I handled chopsticks efficiently and that I had learnt at least two or three understandable words in Chinese pleased my hosts very much.

The secretary-general was far the most commanding personality present. He conducted himself with the greatest dignity, although the fact that he was an agriculturist suggested that he came from peasant stock. He spoke no English, and I do not think he knew the West. This reminded me of the fact that in spite of the sociability of professional men, some of whom are at least partially westernized, any negotiations with the Chinese Government will probably have to be conducted with men of very different character—men with little knowledge of the West and quite probably with distrust and hatred of it. Such men would be likely to have the same pride of race and isolationist point of view that stamps the illiberal politician everywhere.

After dinner I had a brief discussion, as I had been empowered to do, concerning the possible relationship of Chinese geophysicists with the international organizations of their colleagues in other countries. For obvious reasons it was a delicate matter. Geophysicists want exchange of information between all parts of the globe. This has nothing to do with politics, but the two Chinese governments, in Peking and in Taipeh, completely refuse to co-operate. It would be as difficult to deny membership to Formosa as to get on without information from Peking.

Fortunately, no one imagined that I could solve this matter, so I contented myself with emphasizing the genuine interest of geophysicists in all studies of the physical nature of the earth in every country. I felt that the very fact that I had made this visit proved the point, and I made no attempt to seek an immediate solution.

When I sensed that the party was over, I shook hands all round, thanked my host, praised the meal and walked out into the corridor. I stopped by the kitchen and shook hands with the cook who had carved the superb duck. She seemed rather surprised as she wiped her hands on her apron, and perhaps my Chinese hosts were also; but I felt it just as well to remind them that Canadians understand the meaning of a classless society as well as most nations.

The Temple of Heaven

Thursday, August 28th

Next day I started my tour of scientific work by driving to the Institute of Geophysics and Meteorology of the Academia Sinica.

I went to the director's office, and there, over the customary and comforting cup of green tea, I was greeted by the senior scientists: the director, Dr. Chao Chin-chan; the vice director, Dr. Chen Tsung-chi; and Dr. Lee Shan-pang. All spoke English and Dr. Lee had certainly visited the United States. I gathered that the other two had studied in Germany.

It seems that the institute has about a hundred employees of all ranks in Peking, but it directs the work of many others. Dr. Lee, for example, has 150 seismologists working for him in stations scattered about the country. Many of the scientists are college graduates, but it also seems possible for a bright apprentice to become engaged in research and to obtain a bachelor's degree or diploma later. I think that the institute has research students and can award degrees roughly equivalent to Ph.D.'s, but I was never quite clear on this point, perhaps because the policy is changing. Some such institutes in the U.S.S.R. and in China, even though they are not related to universities, certainly do have the power to grant high degrees. For this reason the senior staff members of institutes may be called professors, although they are not at universities. To people with our point of view, all this is a little confusing.

This institute is located in a newly developed region of north-west Peking, in which is concentrated a great collection of new institutes, universities and similar buildings. Over a large region there is tremendous activity to replace the former fields of vegetables with hundreds of new laboratory and office buildings and with dormitories for the workers employed in them. These are laid out along brand-new avenues planted with trees. It seems quite possible that as many as 100,000 scientists and students may be working in that area. The institute that I visited is only a very tiny part of this cultural centre, but its one new permanent building is typical of so many others that I shall describe it.

It is well built of grey brick three stories high, with a tile roof. Few buildings, except hotels, have elevators, so that they are rarely more than four stories in height. Inside there are three stairways connecting central corridors that run the length of the rectangular building. On each floor and in the basement are a couple of dozen rooms. The building is plain but well built, with white plastered walls, terrazzo flooring, electric lights, running water, a few coils for steam heat in winter, and doors and windows that open and shut properly. The senior scientists have bare and simple offices of their own, equipped with the necessary furniture and plenty of books. Laboratories have good benches, blackboards, and sinks, also adequate outlets for electricity, water and gas. These buildings are in fact quite satisfactory for their purpose, but without any frills. I went all over the building, going into most of the laboratories and vaults to inspect projects.

Housed in this fireproof building is an excellent library. The scientists pointed out that they have the responsibility of answering a lot of questions upon which large expenditures may depend; for example, about the construction of dams, bridges or buildings. They need a good library, and they are proud of the fact that over the past thirty years they have built up an excellent one.

By counting the shelves, I estimated that the reading room held about four hundred current journals on geophysics and related subjects. It was impressively complete and up to date, having, for example, four Italian geophysical journals, five Japanese ones, and all the well-known journals published in English, French and German. I also went carefully through the stacks and opened a variety of volumes there. The sets of all important geophysical journals were complete and there were many marginally related publications such as the *Proceedings of the American Society of Civil Engineers,* which would be of value when considering the design of earthquake-resistant structures.

The library and its indexes are all in three parts, for there are large Russian and Chinese sections, as well as that for the Western languages. The Chinese publish about fifty scientific journals at present, but the number is increasing. All may be obtained by subscription. From my own observation all but the most popular ones have at least abstracts, and often whole articles, in Russian or in some Western language.

Although this library is larger than most that deal with only one field, there were good libraries to be found in every institute and university I visited. Back and current numbers of the more widely read journals, which might otherwise be difficult to obtain in China, have been multilithed, and I repeatedly saw copies of some of the standard American, English, German and Russian journals, and of textbooks, in that form.

I was also taken through the workshop where they make some of their own instruments. Crowded into a temporary building were fifty men and women who were operating twenty lathes, a milling machine, a planer, and some other tools, mostly of Chinese manufacture. Each operator was training a young apprentice.

78

The chief mechanic explained to me with the help of Mr. Tien, "We have stopped all regular work for two months in order to make thirty more lathes. We will soon need them and cannot obtain them in any other way."

Pieces of a standard Chinese lathe were being copied all over the shop.

This Institute of Geophysics and Meteorology had been formed, I was told, in 1928 as an Institute of Meteorology with much the same functions as today. It is concerned with meteorology, seismology, geomagnetism, and geophysical prospecting. In 1935 there had been only fifty meteorological stations in China and only two seismological stations. (One was at Shanghai, founded by the Jesuits, and the other was at Nanking.) There are now said to be 1,500 meteorological stations reporting four times daily through local centres, of which four hundred are pilot-balloon stations and seventy are radiosonde stations. The Chinese make daily, three-day, and monthly forecasts by the use of computers, but I did not see the facilities for this work.

They have thirty seismological stations, of which they later showed me three and offered to let me see others. In Moscow the Russians had said that they received bulletins from twenty-three Chinese stations, and that sixteen more were under construction or contemplated. The Chinese also have temporary mobile stations.

In the vaults and scattered around this institute, I saw many Chinese seismographs, of seven types. There are three varieties of the so-called 1951 type, a simple instrument designed by Dr. Lee for the purpose of getting stations started quickly when he was told to do that in 1951. There are small Russian Vegik, standard Russian Kirnos and high-sensitivity Russian Kharin seismographs, all of which are copied and made in China. They have a portable instrument operated from batteries which can be used for preliminary investigations.

The chief job in the seismological division was preparing

a map of China, indicating the frequency and intensity of earthquakes, so that suitable precautions might be taken in building dams, bridges, buildings, railways, and so forth. As is well known, many parts of China (but by no means all parts) have severe earthquakes from time to time. Because only two seismograph stations had been operated for any length of time, the old instrumental records were not complete enough for this purpose. Only in those regions where earthquakes are numerous had there been time for collecting adequate data in the few years since the network of new stations had been installed. By good fortune these data sufficed in the mountainous parts of southern and western China, which had always been thinly settled. For the plains where earthquakes are less frequent but may, nevertheless, be dangerous, Dr. Lee obtained the services of 150 historians who in two years reviewed the whole literature of China and found about 10,000 useful references to individual earthquakes occurring between 1189 B.C. and the present. By careful analysis of these varied and imperfect data, Dr. Lee has prepared the required maps in preliminary form. Two out of the three volumes of his remarkable report have already been published and favourably reviewed in the *Bulletin of the Seismological Society of America.* He showed me his work on the final volume.

It seems appropriate to mention two peculiarities common to all institutes; one concerns physical exercise. In the past, intellectuals in China looked down upon physical work in any form, an attitude that tended to make them impractical and completely divorced them from the labourers. To avoid any continuation of this custom, the present Government is very insistent that everyone engage in both forms of activity. In the middle of the morning and again in the afternoon, at all offices, institutes and colleges, bells ring, loudspeakers are turned on, and everyone troops outside for ten minutes of physical exercises to gay music. In many ways this seems to be an improvement on the coffee break. I always suggested that we go too, but my hosts never allowed this, so I do not know whether the professors normally join in or not. It is taken quite seriously, for I have seen charwomen stop scrubbing and stand up and do ten minutes of exercises before resuming their scrubbing again. For similar reasons there is emphasis on sport, chiefly basketball, for which the nets can be seen all over China. They are so numerous that it seemed to me that it would have been more suitable if the secretary had referred to Canada as the land of Dr. Naismith, born in Ontario and the inventor of the game, rather than as that of Dr. Bethune. In those institutes that have not yet got playgrounds, it is usual for the employees to go out with shovels and baskets and work until they have. Some were doing that at this institute.

The second curious feature is the wall posters which can be seen in varying numbers in the halls of most buildings. These large sheets of white or coloured paper are covered with Chinese writing which, I was told, are statements of criticism or of praise which anyone may put up, directed against the director or anyone else in the establishment. I take it they are frequently detailed, personal and pointed, but I feel it improbable that they are very spontaneous.

In most of the places that I visited, the directors seemed to be chosen for their scientific or technical experience, and I suppose that from time to time some of them may be regarded as less politically mature and correct than they ought to be. Likewise some of the employees may from time to time stand in need of correction. It seemed to me that the wall poster was one ingenious method of keeping technically competent directors, and indeed everyone, on the correct political path without destroying their authority. Another method was interminable political discussions, and I noticed meetings wherever I went.

This particular institute had its share of these posters on the walls and stairways, although they were not nearly so abundant as in some other places. The director admitted with a wry smile that he was the target, but I was interested to observe that his authority had not been lost. As we went around the laboratory, everyone was most polite and deferential. When I later addressed the members of the institute, I could not have asked for a more polite and attentive audience, and when the director and I walked in, everyone stood up. Nevertheless, I often wondered what the dean thought in the university, which was so plastered with these posters that there was no more room on the walls, and posters were hung on strings across the corridors, until it was a little like an obstacle race to visit the laboratories.

These visits occupied all the morning, and I returned to the hotel for lunch alone, where I copied these items from the menu:

Cold Dishes

Chicken in red oil	$ .36
Spicy giblets	.30
Mixed cold noodle with chicken shred	.18
Spices pork	.34
Spices chicken fins	.20
Fried Phoenix fish tails	.32
Black eggs	.08
Vegetarian macedaing	.20

Vegetables

Sauté of rape	.10
Sauté of cauliflower and snow eggs	.22
Sauté of green pumpkin with dry shrimps	.16

Any Kind of Soup

Consommé of spongy bamboo and peas sprout	.14
Consommé of three-deliceous	.26
Consommé of pork and eggs and fungus	.10

Fowls

Mixed soft chicken in piece	.42
Fried spice duck	.72
Fried duck livers	.32

AN ALBUM OF PHOTOGRAPHS

Idle pleasure boats
on a weekday afternoon
at the Summer Palace, Peking.

August in Siberia.
Russian women
near the Chinese Border.

The author beside
the Moscow-Peking Express
of the USSR State Railways.
(This picture was taken
with the author's camera
by Chinese students
on his first morning in China.)

Moving furniture by peditruck to new buildings
in the Academy area, north west of Peking.

A Peking trolleybus, pointed out to the author
as having been made in Shanghai.

Levelling ground for new buildings
outside the walls of Peking.

The gateway to Mongolia
in the Great Wall of China.

Fish and Prawn

Fried Mandarin fish and paunch in sweet and sour
sauce .... .32
Sauté of prawn in small piece .... .34

Meat

Sauté of pork and green pepper in does .... .32
Slice of pork in fish-tasted sauce .... .22

Marin Products

Stewed bich-de-mer with slice pork .... .46
Sauté of fish tripe-shred in snow eggs .... .42

Deserts

Stewed lotus seed in sugar candy .... .12
Almond Bean curd in sugar candy .... .10

Pastry Foods

Noodle in soup with three fresh .... .24
Boiled noodle with hashed meat in bean paste .... .16
Stew vegatarian dumpling (each one) .... .02
Rice congue (each bomb) .... .02

While the occasional misspelled words are amusing, it is more pertinent to reflect that most of the much longer menu from which these excerpts were taken was entirely correct, and to consider how few North Americans could translate even a word, let alone whole menus, into Chinese.

I was reminded of the incident of Dr. A. D. Tshkakaya, a charming seismologist from Tbilisi (Tiflis), Georgia, who came to a convention in Toronto with other scientists from the U.S.S.R. The advance cable listed only the visitors' surnames. Confident that "kaya" is a feminine ending in Russian and not realizing that his name was not Russian but Georgian, we listed him as "Miss Tshkakaya."

When he arrived and complained, a correction was made. Unfortunately, the result was that on the revised list his name appeared as "Mrs. Tshkakaya." Ten months later when I visited him in Tiflis, he and his friends were still laughing over the old gentleman's jest that he had lost his virginity in Toronto.

In spite of the exotic names of the dishes on the menu, I found those I tried simply delicious. It was always a question in my mind whether I should return to some of the delicacies I knew or strike out on a fresh epicurean adventure. For a start, I would recommend to anyone this menu:

Chicken in sweet and sour sauce
Consommé of three-delicious (a clear soup with green
    leaves and pieces of mushroom floating in it)
Pineapple in sugar candy

All the desserts in sugar candy were made by frying pieces of fruit or other sweets for a moment in deep fat and then dipping them into hot sugar toffee. They were brought to the table still sizzling and bubbling from the stove. One picked up the delicious morsels with chopsticks and plunged them into a bowl of cold water to crystallize the toffee before eating them. Of course, I always drank many cups of the mild aromatic Chinese tea, and among the first words of Chinese that I learned were, *"Lai pei cha, ch'ing.* (Bring more tea, please.)"

After lunch I went to my room and wrote out the first lecture I was to give. This was a nuisance, because I had given the same lecture twenty-two times in the United States the previous winter and needed no notes. But the interpreters, although perfectly competent, had asked for them. Through lack of practice they were unsure of themselves, and they wished to have an opportunity to look at the text and particularly at the technical terms that I might use.

At last, at four o'clock, I was free to visit the Temple of Heaven. As we drove through crowded city streets, it was raining and many blue-clad figures were hopping about among the puddles, seeming to hang momentarily from their picturesque paper umbrellas as they leaped. I bought one of these umbrellas for less than a dollar, and it proved to be a most serviceable and strong article. The bamboo frame was no doubt the ancestor of our modern steel version, but for me it had the interest of novelty.

Most of the rickshaw men and some of the other people had on red or yellow oilskin raincoats, and I saw one or two old men wearing raincoats of thatch. The ingeniously laid straws stuck out all over their backs like quills on a porcupine.

The stores along the way were all very small and crowded. While some sold manufactured goods after our style, many were shops of craftsmen who, in extremely cramped space, were turning out tables and chairs, bicycles and pieces of machinery. In others women, crowded elbow to elbow, sewed furiously at machines or assembled manufactured articles.

A drive of two or three miles brought us to a vast and sombre forest, over which the grey skies added to the natural darkness of the cyprus trees. They were uniform and planted in rows. Referring to their most peculiar and beautifully gnarled trunks, Mr. Tien said: "Those are imperial dragon cyprus trees. They are so named because of their curious trunks. This was formerly the royal park of the Temple of Heaven, where the Emperor came to pray."

We stepped out of the car and walked along a sand path into the forest. After a few hundred yards we passed a large number of tents laid out in rows among the trees. They were obviously military, but only a few orderlies were about. Presently we reached a great ramp rearing above us through the forest. We walked up to it, ascended a flight of stone steps, and found ourselves upon a magnificent raised causeway.

This wide and handsome road was without railings, so its grey stone surface blended intimately with the tops of the dragon trees that surrounded it like a sea on every side. It had a gentle slope rising steadily towards, and giving an impression of eminence and grandeur to, the Temple of Heaven standing at its head. This magnificently proportioned shrine was circular in plan and rose in tiers, as though it had been roofed with three giant bells placed one upon the other. Each roof was covered with porcelain tiles of the deepest blue, which glistened in the rain against the dark grey sky. The temple seemed huge and awe-inspiring.

We went up the ramp and through an archway. At our feet lay a great marble bowl, from the centre of which the lovely temple rose in heightened magnificence. Descending into this courtyard, we passed through rows of potted flowering shrubs and climbed the white marble steps to enter the six-hundred-year-old building, regarded by some as the most perfectly proportioned in the world. We marvelled at the vast teak poles and the intricate design of the framework.

As we passed through the archway we mingled with the crowd of people sheltering from the rain, and stopped briefly at the stalls where a great variety of Chinese toys and souvenirs, candy and postcards were for sale. From the number of these, their cheap price and the great throng of people, it was evident that some Chinese people had time to sight-see and a little money to spare. Most of those there that afternoon were women and children, and some of the little ones were amusing themselves by climbing in and out of the bronze urns that served as ornaments around the terraces. I tried to photograph them, but it was too dark where they sheltered from the showers.

I did consider photographing another class of people also there in great numbers, but I decided it wise not to do so. On the sides of the courtyard a battalion of soldiers were sitting sheltering from the rain. While photographing the Temple of Heaven I noticed that when I turned my camera towards the soldiers an orderly was dispatched in my direction, so I surmised that it would be prudent to refrain. Conspicuously and intently, I photographed the central architecture and elaborately put my camera back into its case.

By the time we had finished seeing these things and returned to the head of the ramp, the worst of the shower was over and on the causeway many soldiers were parading to the music of a hidden band. Mr. Tien and I walked slowly down its whole length to another palace a quarter of a mile away. The road was just wide enough for twenty men abreast to march by us on either side as we walked down the centre, brushing past fierce and angry drill sergeants while the columns of troops swung past on either side, only two or three feet away.

"Mr. Tien," I said, "how is it that in your peace-loving and progressive country there are so many soldiers about?"

For a moment he hesitated and then replied, "I believe that they may be a cadet corps from one of the universities undergoing holiday training."

"Congratulations!" I said. "I do not think I have ever seen better-trained cadets."

I thought back to the spring of 1940 to a parade ground at Aldershot, England, where with the help of experienced sergeant-majors we Canadian engineers had desperately sought to prepare ourselves for a general inspection before a trip to France, which never came.

How Mr. Tien could have imagined for a moment that I would believe that these were cadets, I do not know. These men, far from being half-trained youngsters casually parading, were magnificent troops, well drilled and disciplined and practising a precise and exaggerated goose-step in phalanxes of four hundred. It was a tremendous sensation to feel the ring of the stone beneath one as these khaki-clad men stamped past in tight formations twenty wide and twenty deep. They were clearly the *élite* of the Chinese professional army, practising for some ceremonial parade.

Much as I would have liked to do so, I did not take pictures, but walked slowly down the parade admiring the excellent discipline that was being driven home by tough sergeants. The latter looked at me with hatred, but they knew that I was not there by chance and said nothing.

At the far end of the ramp, through another marble gate, we entered the Temple of the Echoing Wall. This beautifully built and ingeniously designed structure was a circular wall, complete but for one small opening.

"Stand close to the wall and listen," said Mr. Tien when we had entered. He went around the courtyard and disappeared behind a group of buildings. Suddenly I was startled by a loud whisper in my ear from the distant and invisible Mr. Tien. I put my face close up to the wall and continued this whispered conversation until, tiring of the amusement, we left the temple and found beyond it a simple raised court. Onto this we climbed by way of three surrounding terraces of white marble, all intricately carved. The central platform was quite flat and paved with nine circles of stone flags about one large central disc. Each circle contained a multiple of nine stones, so that successive rows had nine, eighteen, twenty-seven, and so on up to eighty-one stones.

I stood on the central stone and realized that here for hundreds of years the Sons of Heaven had stood on December 21st to welcome the return of lengthening days. I looked out beyond the symbolic platform and the white terraces to the dark green forest of tree-tops and up the great causeway to the Temple of Heaven, which reared its triple crown in azure blue over the tops of the lesser temples and archways.

Here I was alone among six hundred million people, a stranger in the most populous nation on earth. There were only a few hundred more Europeans in China today than there had been in the days of Marco Polo, and perhaps less western Europeans than in the time of Matteo Ricci and the Jesuits. Certainly there were less Chinese Christians now than then. How odd that science had been the password to the great Forbidden City and to China, both for this long-dead Jesuit and for me!

I thought of the vanished pomp of three thousand years of imperial dynasties; of the ancient and peaceful civilization that had absorbed invaders into itself, and had brought culture, wealth and peace to these vast regions—a civilization that valued scholarship, high principles, art, and good living. For centuries, until this present turmoil, it had scarcely changed.

But I also heard the muffled drum beat, the distant shouts of command, the stamp and ring of military drill upon the temple ramp. At least the Communist régime had not turned its back upon ancient monuments, but had come here with pride to train a new China in the shadow and footsteps of the old. Would this new China follow the pattern of the old? There had been upheavals and dynastic changes before. After the bloodshed and bitterness of revolution passed, would China slip back to the historic form, or was this something extreme and terrifying? I did not know the answer.

In the distance a train whistle blew and a gust of wind sweeping across the platform brought a returning shower. Rain splattered on the stone terraces. I shivered in my summer suit.

"Mr. Tien," I shouted, "we had better get out of here before we get soaked. Let's make a run for it."

Picking our way quickly but carefully down the slippery marble steps, we descended into the forest and ran through the dark, dripping dragon trees to the waiting shelter of our car.

We drove through the rain back to Embroidery Street and another shop much like the first. This time I was not the only customer, for an East German and his blonde wife were engaged in purchasing a fur coat.

I spent so long there that we had no time for supper. Mr. Tien suggested it, but I said I would rather go to the opera than eat. So as we drove, we dined on one of the chocolate bars that I had carried with me from Europe for such eventualities.

Mr. Tien was not very sure where the theatre was; and to reach it, we abandoned our car and walked some distance down the narrow streets. They were quite clean, thronged with people, and lined with shops in which work was continuing at full speed by very inadequate electric lights. With some difficulty Mr. Tien located the opera, which was no larger or more conspicuous than a small movie house. We went in.

The outside of the building, which was probably an old one, was too dark for me to see or remember; but when we entered the theatre lobby, I found that it was well lit and had light buff walls. Attendants at tables were selling programs and soft drinks. We hurried through and entered a darker auditorium just as the performance was about to begin.

I am not sure of the details of construction of the theatre, but it left me with an impression of bare concrete floor and walls. The roof was decorated by hanging curtains of woven bamboo, which would certainly constitute a major fire hazard. The seating was arranged in conventional fashion, with rows of bare wooden benches in front of a stage and orchestra pit. Our places were in the centre of the third or fourth row, so that when we had wormed our way in among the blue-clad Chinese who filled the theatre, we had an excellent view of the performance.

Gongs were already beating in the orchestra pit, but I had time to look around and see that the Chinese were all dressed in ordinary clothes. There was no way of telling whether they were a typical cross-section of the population or not. Certainly they represented all ages, and close to me a mother held on her knee a child clad only in diapers. During the intermission she took it home. There were lots of other children, who were well behaved, and the audience was like that in a movie house.

With a crescendo in the banging of gongs, which were to keep up a ceaseless din throughout the evening, the show began.

There was no orchestra such as we are accustomed to, but instead a variety of metal and wooden cymbals, gongs, and sounding boards. I think that there were drums too, but the gongs made most of the noise. These maintained a general background of rhythm, which rose to a deafening clanging when the action became exciting. Whenever the hero stamped his foot and scored a point, whenever the villain entered or embarked upon some nefarious plan; whenever the poor heroine, after much lamenting and wringing of hands, finally made up her mind (usually the wrong way), the gongs beat furiously to emphasize the drama of the moment.

So far from being an opera in our sense of the word, the show was a first-class melodrama in which songs played a comparatively minor part. It was brilliantly acted, gaily costumed, and enriched by clowning, tumbling, singing, and especially the exotic gongs.

I found it splendid entertainment and easy to follow because of the superb acting. There is no question but that all the performers were professionals with many years of training and great skill. The costumes were elaborate and changed with every scene. The make-up was far more fantastic and exaggerated than anything I could have imagined. There was little scenery and props were comparatively simple; lighting was used to differentiate the dark forest from well-lit indoor scenes.

The play, which was called *P'an chin liu,* had been written a thousand years ago and was still played in its traditional form from the Sung dynasty. The plot would be recognized by anyone familiar with seventeenth-century European drama, and contained all the stock figures—the doleful cuckold of a husband, the flirtatious and silly wife, the evil procuress, the handsome villain, and the upright and avenging brother. It pursued the fortunes of a poor porter and his young and beautiful wife and came to a roaring conclusion with evil lying dead and bleeding all over the stage.

The porter's clothes were ragged, his face was dirty, he was stooped and bent, and he was generally bathed in floods of tears, while the heroine simpered about the stage with little mincing footsteps. She presented a most attractive appearance, for her graceful figure was clad in an ever-changing succession of fabulously rich silken gowns. The only part of her costume that did not change was a vast structure surrounding her head and face that reminded me of an elaborately decorated and bejewelled bird-cage. This formed an effective frame for her face, whose downcast and modest glance did not conceal her small and pouting mouth or her flirtatious eyes.

Her fine dark eyes were emphasized by heavy, almost crimson, make-up that extended from her eyebrows down over her lids to fade gradually away on her cheeks. Her nose was white, but along the line of her eyebrows the colours were reversed, so that a central, vertical strip of pink colour extended up from her nose to divide her forehead, the sides of which were white. Strange as this make-up may sound to us, it was nevertheless effective and she was a very beautiful coquette.

The tedium of their bickerings was relieved by a pair of clowns. I did not know what they were talking about, although they convulsed the audience, but I was fascinated by their appearance. One had a large circle of white covering all the centre of his otherwise brown face, from his upper lip to his forehead and out to the middle of his eyes, which gave him a very queer effect indeed. The other had vertical white strokes through his eyes, so that they looked like car headlights in the wartime blackout in England or like characters in ancient comic strips who had plus signs for eyes.

The villain was clad in mandarin coats richly embroidered and generally light in colour, and wore on his head a curious hat with a cockade on a spring that wobbled back and forth in a most disconcerting fashion whenever he moved. Incongruously enough, the hero wore a close approximation to black velvet doublet and hose, until the final triumphant scenes, when he strutted about in a dark mandarin coat whose sleeves had pockets hanging to the ground like a stole. The ends of these kicked forward into the air with every stride. It was a most effective gesture of supremacy.

Of course, the hero had his difficulties. In one exciting scene, when walking unarmed through the forest, he met a tiger. On the half-darkened stage a tremendous fight ensued which called for great acrobatic skill on the part of both the actors. Repeatedly the tiger and man rushed at each other or leaped over one another and rolled somersaulting across the stage. The gongs beat wildly. It was very exciting action and of course the hero won.

While the hero was busy with the tiger, the villain slyly made his rendezvous. He led the heroine on a little excursion off the stage. Secretly the old woman delivered poison. Smoothly the heroine administered a dose in the simple porter's supper. In agony and convulsions on the stage he died. Too late the vengeful hero returned, fierce, formidable and bearing a vast two-handed sword.

He trapped the villain and a terrific fight ensued. The sword was whirled so rapidly, first by one actor and then by the other, that one did not notice whether there was one sword or two. It twirled continuously, now round and round at waist height with the intended victim bobbing up and down between every revolution, then high overhead and flashing down first on one side and then on the other. It is a miracle that both of the men did not get decapitated and quartered.

A crowd gathered to watch vengeance exacted. The hero, when he had chopped up the young man with the cockade, sliced up the wicked old woman as well, and the heroine seized the sword and fell upon it.

With every gong beating furiously and the crowd roaring its approval, the curtains swept to the floor, and we followed the Chinese audience into the night.

Science and culture among the cabbages

At 9:30 I went to the Institute of Geophysics and Meteor-ology to give my first lecture. The director received me graciously and provided a cup of tea before leading a few senior scientists and myself up the stairs to lecture. As we entered all the students stood up respectfully, clapped their hands and waited politely till we were seated. The lecture hall was a large classroom.

From ten o'clock until nearly twelve, with Dr. Chow Kai interpreting each phrase and sentence as I went along, I outlined my views on the physics of mountain building to an audience that remained quiet and intent although, due to translation, the lecture took twice as long as it otherwise would. In spite of the fact that I had no slides and therefore had to point to maps and take time to draw diagrams on the blackboard, the audience sat absolutely still and attentive.

These young people might post notices on the walls sug-gesting how the Director should run the Institute, but when he had guests they were certainly well disciplined and knew how to behave. I felt they were genuinely keen and anxious to follow a fresh point of view.

After a few questions by senior members of the staff and another cup of tea we drove back to the hotel, dropping one or two of the scientists for lunch at their dormitories. This was the term they invariably used for their quarters. They never spoke of homes, houses, flats or apartments. These were crowded but clean new buildings in which they lived with their families, for we saw women and children and they spoke of them. I gathered that at best they were lucky to get three small rooms for a family.

We had time for only a hasty lunch before we picked them up again and drove out of the city along paved roads, past new factories and institutes, into the Chinese countryside. Peking was clean. Every street and yard was swept every day, as one could see each morning. So was the country. There were no dogs, mice, flies or mosquitos whatsoever, and very few cats.

Although the people were often poorly dressed in faded cotton clothes, very few of them were ragged except some of the men engaged at the hardest physical work of driving animals, pulling carts, or pushing pedicabs. Blue was the commonest colour but others wore white, black or khaki, and occasionally one would see a girl in a pink or green print blouse. Men and women alike were dressed in Western-type shirts and trousers. Mostly they wore their shirts outside the trousers and the girls' shirts were cut a little shorter and a little differently at the neck from the men's. Sometimes they wore a cotton jacket or carried an umbrella. Only a few of the old people still had more Oriental clothing, but it too was simple.

All of the men had their hair cut in Western style. They were clean-shaven except for a few old gentlemen with picturesque and straggling beards. The girls had their hair cut in a neat bob, often with a bow in it; the older women swept their long hair back into a tight bun.

It was sad to see many of the old women limping along on tiny feet that had been bound in their youth and were

95

so small that they looked like peg legs. This practice has of course completely vanished. In the country around Peking I noticed that everyone had on shoes—cotton ones with rope soles—except one party of men who were sitting beside the road resting, who had apparently taken theirs off to cool their feet.

When we had driven a few miles outside the city we were stopped by a barrier across the road and two or three policemen, who examined the pass my companions carried. They allowed us to proceed and we soon reached the foothills where the Central Geophysical Observatory of China was located on bedrock in quiet country surroundings. A high brick wall surrounded the compound, in which were the eight or ten buildings of the observatory — three magnetic huts, a solar observatory, cosmic-ray and seismological buildings, and two or three dormitories.

On that beautiful afternoon it was very pleasant to make our way to the various buildings through the well-kept vegetable gardens and flower beds, with their backdrop of soft green hills. Across the middle of the compound there was a sharp, narrow ditch that no vehicles could cross and that reminded pedestrians that they were entering the nonmagnetic area and should not carry iron objects that might disturb the delicate instruments. Having divested ourselves of such things as keys and pocket knives, we entered and examined the three magnetic observatory buildings, one each for recording, for calibration, and for experiment. They had a bare minimum of geomagnetic instruments, but room for many more. We also went into the solar observatory and I looked at the afternoon sun through their large telescope, with which two young men were watching for solar flares and photographing the corona as part of the program of the International Geophysical Year.

Although China had withdrawn from official participation in the IGY program when Taiwan was admitted to it, the Chinese told me that they were continuing much the same work, which they had planned and started, but they were not transmitting the results to the World Data Centres of the IGY. Another building, for the study of cosmic rays, was not operating. The upper atmosphere was being studied elsewhere.

"Long before the revolution," Dr. Lee said, "I started the study of earthquakes in this vicinity at a place some ten miles farther from Peking, called Chufan. In those days there was no highway and I had to work almost alone. It was difficult and discouraging. Since the liberation progress has been rapid, and in 1955 the geomagnetic station was opened. We had great difficulty in constructing the underground vaults for observing earthquakes, but the seismic work was transferred here last year."

The vaults are now complete, and I went down three flights of stairs into them and examined three sets each of three instruments of Kharin, Kirnos and Galitzin types, all copied by the Chinese from Russian models. They also showed me one electronic seismograph of an experimental type and two mechanical recording instruments of very low sensitivity. These would give them records of any strong earthquakes, so large that they would break the regular instruments or go off scale.

There was only one young man to receive us and provide tea. "Five seismologists work at this station," Dr. Lee said, "but today they are in Peking to attend planning meetings. This is the central station for all China, where the seismograms from all twenty-three stations are sent for examination. Of these twenty-three stations we consider thirteen to be first class, with both Chinese Kirnos and Kharin instruments. The others have only mechanical registration for large earthquakes. We hope gradually to add eighteen permanent stations, but in the meantime we have a number of mobile stations searching for the best sites." This confirmed the information I had already heard in Moscow.

He thought the Kirnos instruments were good but the

Kharins were useful for only some types of earthquakes. He told me that when they had installed these stations they had expected to record many small local shocks, but they did not find them; or rather, they occurred only sporadically, in showers. At one time Peking had been notorious for its large earthquakes but there had been none of these since 1730, showing how the seismically dangerous areas migrate about the earth with time.

Then he told me there were still a seismically active zone north-east of Peking, extending to the coast, and another to the north-west of Peking, very active and continuing to Sian and Szechwan.

"In my early days, during the revolution and invasion and before the liberation, work was very difficult," said Dr. Lee. "In all China only one or two others besides myself had any training, but now I have 150 seismologists and many students working with me. Some of the stations will have only one or two trained persons, but they are going to get more.

"Every station prepares preliminary reports which are sent by telegraphic code to the Institute. From these reports we compile and issue preliminary bulletins, and later final bulletins when we have a chance to re-examine all the records.

"Within eight hours after every large earthquake we give the Government reports upon its location and intensity. In 1954 there were five destructive earthquakes and several in the two succeeding years, but in 1957 and 1958 none of great consequence.

"We still grade earthquakes into twelve classes, depending upon the degree of damage," he continued. "For example, the last serious earthquake in China was of Grade 10 in Kansu in 1954, but I prefer the new system of measuring magnitude that is being developed in the United States. Eventually we may adopt it."

After this very thorough examination of the station and

another cup of tea, we drove back to Peking through the peaceful late afternoon. Most of the road was paved and young trees had been planted along it. "The Japanese cut down all the former avenues of trees," said my hosts, but they pointed out that the avenues had been replanted with fast-growing trees that soon gave shade, alternating with slower-growing trees for permanent protection.

There was not much traffic on this back road, and most of what there was was drawn by horses, pulled by men, or pedaled on bicycles. Beside the road the Chinese worked their fertile fields, or sat around in the villages and watched the children play.

Back at the hotel I had only time for a hasty supper, which was a pity because it was such a good one: chicken with sweet and sour sauce, pumpkin with dried shrimps, and pieces of apple fried in deep fat and covered with hot toffee.

Mr. Tien and I raced off to a theatre larger and more modern than the opera house of the night before. We sat in the front row of the balcony to watch an exhibition of folk dancing. As before, the theatre was crowded with people and half a dozen little girls about seven years old sat immediately behind me and squealed with delight throughout the performance. It was a cross between the Moisiev dances and the type of dancing put on in Toronto every year in the grandstand show at the Canadian National Exhibition. I was interested to observe the very strong current of similarity between this Chinese troupe, the Moisiev dancers, and the folk dances that I had seen in Romania.

The names of the dances were different but the program and even some of the gay costumes were similar. In Romania they had danced Moldavian, Bessarabian, Bulgarian, and Transylvanian dances. The Moisiev troupe specialized in Ukrainian, Uzbek, Turkmanian and Polish dances, while the Chinese did Tibetan, Mongolian and Sinkiang dances. In every case the emphasis was placed on the dances and songs of the national minorities; but, perhaps due to my lack of familiarity with folk dancing, they all seemed to be very much the same to me. All of the dancers were extremely well trained. They were young and vigorous and alternated tremendous leaps with very fast and skilful footwork. It was a great pleasure to watch them.

In China they varied the folk dancing by singing plaintive mountain songs from the border lands. They also introduced large elements of propaganda. For example, they had a dance of plenty around a vast grain sack like those to be seen on posters all over the country celebrating the bounty of the 1958 harvest. The dancers cavorted around this, tossing huge papier-mâché cabbages and turnips into the air in their delight. There was a dance celebrating The Great Leap Forward, organized rather like a Kentucky minstrel show with end-men playing accordions. And the girls did a very beautiful dance in which each played the part of a petal in an unfolding flower. "This dance," Mr. Tien said, "illustrates the unity of the minority nations."

The final number brought down the house. It was a traditional Chinese lion dance, the old vaudeville horse act raised to the nth power. Each lion consisted of two men concealed in yellow fur, with a vast papier-mâché head, a mane of flowing golden curls, a collar of jingling bells, and a long tail behind. There were a dozen adult lions, as well as several cubs that were each played by a single man — also two lion-tamers in cavalry costumes. All that I remember of them was their great whips, which they cracked continuously.

The plot escapes me now, but it was a mêlée of skilful tumbling, of growling and fighting, of organized confusion and riotous motion to an accompanying cracking of whips, tinkling of bells, and banging of gongs. The high point in the by-play was the dropping of a cub by one of the lionesses.

The Great Wall

Saturday, August 30th

For the benefit of the interpreter the next day I had to write out my second speech, which I did as fast as I could after breakfast. By 11.30 it was done and Dr. Chow Kai, Mr. Tien and I were free to visit the Great Wall. We drove out of the city on the road to Manchuria, across the fertile Peking plain, and climbed a valley into the hills beside a railway until it disappeared into a tunnel. Eagerly I looked for the wall, but the bare hills pressed closer upon us, the road became narrower and steeper. I could see nothing but road cuts beside us and the gorge below. We were almost up to the pass before a guardian gateway reared before us, with the ancient wall winding up the highest ridge on either side of the wall to its top. There I saw that a pathway wide, in good repair and looking surprisingly trim and new for all its 2,300 years.

At the pass the road ran through a fort on the Great Wall. We drove into it and parked the car in a space surrounded by this ancient monument. A flight of stone steps led up the side of the wall to its top. There I saw that a pathway guarded by a crenellated parapet on either side surmounted the wall for all its length.

The wall ran on over green and jagged ridges as far as the eye could reach, on and on over the hills. At intervals of a few hundred feet it broadened into a square turret — a simple room on the wall, now bare and roofless, but having a higher parapet and windows instead of battlements. Perhaps every tenth turret took the form of a larger two-storied fort. In my imagination I could see it cresting the mountains for 1,500 miles to the rim of the awesome Gobi desert — the symbolic frontier and limit of Chinese imperial sway for two millennia.

It was already 1.30. We sat down in the shade of the nearest wall to eat the hard-boiled eggs, rolls and pieces of cold chicken and ham that constituted our lunch, and drink a bottle of orangeade. The cardboard boxes in which our lunches were packed were hopefully papered with a design of Picasso's doves of peace. It seemed like irony to me, and my optimism about the prospect for peace was tempered by the mimeographed sheet quoting the BBC news that I had received the previous evening from the British Embassy. It told me for the first time of the Quemoy shelling.

Immediately after our meal, although it was quite warm and rather hard work, we walked along the wall to the west, climbing steadily with it for a mile or so to the top of the nearest hill. On the way we met a dozen other tourists coming down. Some were Chinese and some were other Asians. Scribbled on all the wall were the names of people from every country in the world.

From the top of the hill the view was even more impressive. Our vantage point was one of the larger forts that bastioned the wall at intervals. Looking south through the hills and valleys, I could catch glimpses of the plain in which lay Peking. To the north successive valleys and ranges broke like waves across the road to Manchuria. In the valley below an occasional car and the much more frequent carts crept up to the fort, passed through it, and crossed over. Occasionally a railway train puffed diligently up the grades, disappeared into the tunnel, and presently emerged to run faster down the opposite slope.

It was rugged, beautiful country, a welter of steep hills and ridges, deeply eroded, but now bright green and rich with grass. It appeared that grazing had been stopped. Not an animal was in sight and the vegetation was coming back.

We had gone farther than most tourists and had reached the end of the part of the wall that had been repaired. Where it continued it was still upright and very much all there, but the balustrades had crumbled and occasionally a piece of the wall had fallen away and slumped down one of the steeper gullies. The wall was a truly magnificent feat of building, for it always followed the tops of the ridges up the steepest places. The bulk of the wall was of rubble, but all of it had been faced with blocks of local granite and paved on top with large tiles. These had also been used for balustrades.

While we were at this fort a shower passed. We saw the grey curtain of mist pass over the clear-cut emerald hills and the wind beat the grass into a silver setting around it, but for all its beauty the rain was not welcome. We got pretty wet and it made our descent down the steep slopes and steps even more dangerous. The wet tiles on the precipitous slopes were slippery and treacherous. I did not envy the soldiers who in the past had kept sentry duty on these steep places. On rainy nights and in the winter snow, they must often have fallen and tumbled a considerable distance.

One of the Australian Communists at the hotel had his hands bandaged from having done just that, but I had little sympathy with him, because I had heard him telling the Chinese at the hotel in a loud voice of the evils of Australia. It seems most unfortunate that a majority of the people who visit China have little good to say about the Western world. This emphasizes the hate campaign against the West, which is strong enough in all conscience. It also occurred to me that the opposite was true. Most of what we have read in North America for the past ten years about the new régime has been wholly bad. Without subscribing to its doctrine, I could not help feeling that a lot of the things being done in China had needed doing and were being done well. The country was cleaner, healthier, better fed than I had expected. There seemed to be a genuine feeling of enthusiasm and hope.

The wall itself was very simple, but somehow its grandeur, the thought of its antiquity, and the views across these wild hills held us, and it was some time before we returned to the car.

In the corner of the parking lot Dr. Chow Kai said, "Look at the old guns." He pointed to a pile of astonishingly primitive bronze cannon. They were small, thick metal tubes, no doubt many centuries old.

"They must have collected them while they were repairing the wall," he said.

"Please come!" said Mr. Tien.

"Do you mind if I keep this chip of tile?" I asked as I put it in my pocket.

We drove back by the Valley of the Tombs of the thirteen Ming emperors — a nearby place among these hills, remarkable for its natural beauty and peace. As we approached the tombs the road became an overgrown boulevard lined with strange marble beasts. They were Ming sculptures, looking like grotesque figures from a medieval bestiary. We visited the tomb of the first important emperor, though I believe that he had had two short-lived predecessors.

There was a park at the entrance where people were sitting and resting in the shade of a grove of pine trees. We had a bottle of beer at the restaurant, and for a small sum entered the crimson walls to walk through courtyards leading to the tomb. These were like those of the Forbidden City with which they were contemporary, but on a smaller scale and the colours were more vivid. We passed through archways of scarlet wooden poles joined by intricately painted screens in blue and gold and green and topped with bell-cast roofs of imperial-yellow tiles. The courts between were small and full of beds of flowers, which gardeners were tending. Beyond these first courts was a vast pavillion supported by more than forty huge teakwood columns. Each was a single unpainted log more than three feet in diameter and of tremendous height.

"These logs were dragged by men all the way from Yunnan three hundred years ago," said Mr. Tien. I felt from the way he spoke that Yunnan might lie beyond the moon. "That was the way the emperors treated our people. This is the building in which the nobles came to worship the emperor and pay respect to his memory. The tomb is beyond."

We walked through this temple as in a forest of giant sequoias, and around a screen at last beheld, across a stone courtyard lined with pine trees, the ultimate tribute subjects could pay to their emperor — or more probably the ultimate tribute the emperor could spend his reign in erecting to himself — a massive, crimson fort that rose to an open, pillared balcony below the eaves of a bell-shaped roof.

We entered this fort through a subterranean tunnel and climbed inside to the upper balcony, from which we had a superb view over the Valley of the Tombs. Directly beneath the vaulted roof, in an open room, stood the tombstone — a single vast block of pink marble, hauled here from Yunnan by slaves. On it was inscribed the insignia of the emperor. It was beautiful and impressive in the slanting sunlight.

"The emperor is not buried here, you know," said Mr. Tien. "They were afraid of grave robbers."

"Where is he buried then?"

"They think that his body is probably hidden in a cave in the hills behind the tomb. Archaeologists are looking for it now."

From the tombs themselves it's not a long drive to the

Ming tombs reservoir, the high point of the sight-seeing tours of every visitor to China for the past year. Many have recorded seeing tens of thousands of men and women carrying baskets of soil to build this great earth dam. The labourers were reported to be volunteers.

The dam was now complete, both sides were faced with stone, and a few workmen were building a balustrade along the top. Lamp poles had been placed in position along it and each one of them bore two flying horses, the Greek Pegasus, that supported the lamp brackets. This, of course, was the symbol of the Great Leap Forward of 1958. Water had begun to accumulate on the up-stream side and to make a lovely lake, but down-stream the land look scarified and barren where the hills had been scraped bare to provide earth for the dam.

A few battalions of men in khaki uniforms, commanded by military officers — I think it probable that they were all soldiers — were carrying out the skilled work of pouring concrete to complete the spillways on the dam.

We drove through the much-reduced camp of tents and huts built of bamboo matting, and noticed that some of the officers had energy enough after a day's work to be doing gymnastics on horizontal bars. Others were getting bowls of rice from kitchens. Beside the road there were piled tens of thousands of broken wicker baskets — mute testimony to the method of construction.

That night, after dinner, as I was returning to my room I ran into another Canadian, Ted Allan of Toronto. He told me that he was a writer and that one of his plays was running in London. He suggested that I join him at supper and meet Sydney Gordon, who was also staying in the hotel with his wife and daughter. Together he and Gordon had written a book on Dr. Bethune, and they were now employed by the Chinese Government to rewrite it as a script for a motion picture that a French company was to produce.

I was glad to have a chance to talk to fellow-countrymen

and I welcomed the discovery that these two Westerners were being paid to do their regular job as playwrights and authors in a cause that might do something to remove some of the prejudices and barriers existing between the Canadian and Chinese peoples.

I found the Chinese whom I met to be warm and intelligent people, humans exactly like ourselves. The bitter barriers of enmity between nations could do no one good; they would solve no problems and they might easily lead to war. Exchanges of ideas and goods seem a prerequisite to better understanding and the opportunity of living out our lives in peace.

Another reason for my pleasure at meeting the Gordons was that I needed some expert feminine advice, for I had been bewildered by all the silks and furs in the shops and knew little of their value. When I explained my dilemma to Mrs. Gordon and her daughter, they assured me that they had little to occupy their days, and would be glad to visit the shop and give me an opinion on the merits of my selection.

In the same room was a party of Canadian lawyers, representatives of the Canadian Bar Association, who had been invited to inspect Chinese courts and legal practices; some Australian trade unionists, and Rewi Ally, a well-known New Zealander, who has spent most of his life in China. He was named after a famous Maori chief. I was interested, some months later, to find a bust of him in a place of honour in the museum in Christchurch, New Zealand. It was in a room devoted to Chinese art; and almost all the room's contents had been sent by Rewi Ally.

Skulduggery

Sunday, August 31st

It was the delightful weather of early autumn, warm in the middle of the day, but cool and equable at night. Each evening when I got back to the hotel I was weary, but I found the tonic of a hot bath and the comfort of a good bed to be wonderful restoratives. After a summer of comings and goings I enjoyed the relaxation of being in one place for ten days.

Each morning on awakening I looked out to see the brightening sunlight dispel the early-morning mist over the low tiled roofs of the ancient city. Beyond them rose Coal Hill, an artificial mound set with fanciful palaces and curious trees to please a forgotten emperor. Peking looked like the illustration in a fairy story and my journey seemed to me like a fable. Once more I felt like a child reading Marco Polo. Sometimes I literally pinched myself to make sure that it was real and that I had indeed crossed Asia by the Trans-Siberian Railway, that I had reached Peking and that I had every expectation of flying safely home from the Orient to my family in another three weeks.

The service in the hotel was excellent, and the room servants, who understood a few words of English, were courteous and obliging. They automatically shined my shoes, and would take my laundry in the morning and return it, neatly ironed and folded, the same evening. To get my clothes cleaned and pressed, which they did for a small sum, I had had to unpack nearly everything and I did not bother to put very much back. Every morning they made up my room, but as they did not disturb my belongings it steadily became more and more untidy.

As soon as I arrived I unrolled the bundle of geological maps that I had brought from Russia and laid them out on the second bed. As I studied the six large sheets of the Russian geological map of Eurasia, I considered where I had been and where I was yet to go, and I tried to learn the geological framework of China. To anyone interested in the study of world geology, travels such as this are of an enormous advantage. They provide a picture of countries upon which later reading can be more easily fitted together, like placing the pieces of a jig-saw upon a copy of the puzzle instead of upon a blank surface.

My suitcase and a briefcase had been bulging with papers and reports, souvenirs and toys from Russia, and pieces of rock that I had collected. I had had to carry all these things across Asia because of the difficulty of mailing parcels in Russia. The regulations permit one to mail papers and books from any hotel, but in Moscow all other items, such as souvenirs, can be posted only at the central post office, which I had had no time to visit. I was glad to find that in Peking I could do up and mail without question any parcels I wished. Parcels and letters travelled quickly to and from Canada without any evident censorship.

I never tried to hide anything or to lock my bags. I never detected anyone investigating them, but if any searches were made they would undoubtedly have been executed skilfully and I might not have noticed. It was not a matter which concerned me for I had nothing to hide except my visa and a few letters concerning Taiwan, and these I always carried in my pocket.

109

It was George Towers, an American who spent much of his life in South America, who pointed out to me the wisdom of travelling with old bags without hotel stickers.

"Such luggage," he had said to me, "is rarely stolen and does not invite the attention of customs officers."

I had also learned the advantages of travelling light. Carrying a battered old suitcase, a briefcase and a light coat, I was self-propelled. Wearing rumpled clothes and assuming as lost and vague an expression as I could muster, I greatly enjoyed playing the role of an absent-minded professor strayed from his campus. My limited baggage created no suspicion, my nondescript appearance excited no envy. I played a role I enjoyed and for which I felt well cast. It proved to be an exceedingly easy way to travel and gave me the minimum of difficulties.

People have often asked whether I was followed or whether I was free to go where I wanted. There was never any need for anyone to follow me. On the train from Moscow to Peking I'm sure that no one shadowed me. So long as I stayed with the train, why should anyone worry if I wandered up and down it or walked on the station platforms or talked to people in the dining-car, as well as I could without any common language? If I had been so foolish as to get left behind at some station it would have been very quickly noticed. The gate-keeper at the entrance to the station would have stopped me and trouble would have ensued.

In Peking I got up and I went to bed when I liked, but of course the boys on the floor knew when I was in my room and when I was out. If I had come in very late or gone out in the middle of the night I suppose their suspicions would have been aroused, but I did not do these things. Each afternoon or evening when Mr. Tien and I parted for the night we would arrange the time to meet next morning, in the lobby of the hotel or in my room. When he was not there I was perfectly free to wander around the hotel, which was a huge rambling place, to visit the rooms of Canadians or other guests whom I met, or to go to any of the shops or restaurants within it.

If I went outside the door, nobody stopped me. As a matter of fact the program was so busy that I had very little time to do this, and I only once or twice went beyond sight of the doorman. On one occasion I went to the Marco Polo antique shop nearby, where the proprietor spoke English. On another evening I went to dine at the British Embassy a few blocks from the hotel. It was a wet night and the streets were deserted. I looked behind me and I was sure that no one walked after me, but I suppose that the times of my arrival and departure at the Embassy were reported by the policeman on duty there.

The truth is I had no means of checking how carefully the movements of foreigners were controlled, and I was very anxious not to try to find out. To have created the impression of snooping would not only have been false, it would have been foolish. It might have prejudiced my chances of seeing those things that I wanted to see and it might have injured my relations with the Chinese scientists. I had come to see Chinese geophysics and I never forgot that that was my objective.

Of course the suspicious can rightly say that on all my journeys from the time of my arrival in Peking until I reached Hong Kong I was accompanied by Mr. Tien, but on the other hand I don't know how I could have got around without him and seen what I wanted to see. As the reader will discover, my route through China after I left Peking took me to the places that I asked to see and not to the cities that my hosts had originally suggested I should visit.

The fact of the matter is that while I did not feel noticeably constrained, neither did I feel entirely free. I was taken on a conducted tour, but it was the tour for which I had asked and it was specially and intelligently designed to show me as much of what I wanted to see as was possible in the time available. I got the general impression that the Chin-

111

ese were exceedingly proud of what they are accomplishing and wished to show the best of it to visitors. Foreigners are few and very conspicuous and are easily kept track of.

On this particular day, a beautiful Sunday morning, it was arranged that I should visit the caves of Peking Man, a famous archeological discovery of interest to any geologist and especially to me, for it had been made by Dr. Davidson Black, a medical missionary and fellow-Canadian. At nine o'clock Mr. Tien and Chow Kai and I got into our Pobeda and drove southwest across the plain of Peking. It was a pleasant drive of perhaps thirty miles to the hills, and as before along both sides of the paved road a double row of young trees had been planted.

All the way there were farms, and at one point I saw a group of people walking through the fields in a long straggling line. One of them carried a red banner, so I suppose they were collectives belonging to some commune; but I remember very clearly that the village they were leaving had not been torn down and rebuilt in barracks. It was an old village with mud houses, haystacks, disintegrating mud walls, and a few old trees for shade, set in a garden-like pattern of fields.

At one place on this drive brick buildings were being erected for what appeared to be a new collective farm and several red-brick factories were under construction, but these stood out conspicuously. I watched China from train windows for over three thousand miles, and most of the villages I saw were of mud and centuries old. It was indeed impossible for the Chinese to have taken down and rebuilt all the villages in China, even if they had wanted to do so. If one considers the problem of trying to modernize a country so desperately poor as China, one can understand that even the greatest wish to do so could not lead to the rebuilding in one year of all the houses for a population three times that of the United States. On the contrary, the Chinese were putting their greatest efforts into building an industrial economy.

Seeing the collective farm, Mr. Tien became quite excited and said, "That is to be one of the communes. Great progress is being made this summer in their establishment."

At first I scarcely understood these remarks, for it must be remembered that I had left Canada in July before the communes had attracted attention. Seeing this group of new brick buildings, I asked him whether they were replacing the old villages.

"Of course," he replied. "It is part of the program of the Government to replace the old mud houses with new brick ones, but it is a large task. They are taking the old mud houses away and spreading the clay on the fields."

"What on earth is the point of doing that? Surely the clay is of no use on the fields?"

"On the contrary. The mud of any house that has been inhabited for more than twenty years is so dirty that it is useful for fertilizer. Thus one kills two birds with one stone, getting rid of the old houses and improving the land. The Government hopes to get all the people into brick houses within five years. Mud houses are not fit for human beings."

During the time I was there I did not see any married people living in barracks, but I did see a great many houses, old and new, in which families appeared to be living a normal life. The program of establishing communes, as I saw it, meant that people were being organized into gangs for work, that villages and factories had community kitchens, dining-rooms, wash-houses, and day nurseries. The program was designed to increase efficiency and to improve standards of living, and appeared to be doing so. What is being done in China must be considered in terms of whether it is an improvement over the former extremes of poverty and backwardness, rather than whether the conditions are as good as those we are accustomed to in the West.

Presently we crossed a nearly dry river by the Marco Polo Bridge. Evidently the river became swollen at times by flash floods, for there were wide banks of gravel on either side. Walls marked the limits of the dry bed that this ancient stone bridge crossed. The posts of the balustrades were carved with grotesque figures of lions and dragons and at one end were a number of plump Ming beasts, camels and cattle, carved life size from white marble. This bridge is famous because it was mentioned by Marco Polo and because the incident marking the outbreak of the Sino-Japanese war occurred there.

Wading in the shallow pools along the river were so many fishermen that I was surprised the river was not fished out. They wore big hats, and cast nets.

Occasionally in the fields were monuments or graves, but in the Peking area these were few. I saw only one cemetery near Peking, and I suspected that most of the ancestral tombs had been dug up and the land put to the plough.

When we arrived at the foothills we turned off through a small town of which the chief industries were quarrying and mining. The hills around were scarred with quarries, and no doubt it was in the course of this work that the caves of the Peking Man had been discovered. Lime-burning was still a major industry, and donkeys with panniers full of coal were being driven along the streets to the kilns.

Beyond the town and a little way up the hillside, we stopped in a modern parking lot and walked by pleasant stairs and paths up through a grove of trees to three new buildings of grey brick, one story high. At the doorway of the first we were greeted by a middle-aged scientist, Dr. Lam Po-chia, and his diminutive girl assistant. She looked little more than fourteen, but I was told that she had a Ph.D. in archaeology and was in immediate charge of part of the work.

We all went inside and sat down at a long table in a waiting-room that also contained a small library. On the shelves were bound copies of the original reports by Dr. Davidson Black. There over a cup of tea the scientists explained their program.

114

Soldiers
pouring concrete for the spillway,
Ming Tombs Dam, Peking.

Billboard outside a historic
Buddhist temple, Sian. The poster
is for the "each one teach one"
literacy campaign.

Faculty dormitories
for six thousand students
at the College of Prospecting, Peking.

Boys on the street in Peking. They are well-fed,
well-dressed and healthy, as everywhere in China.

Top: Chinese tourists in the gardens of the Forbidden City, Peking.

Bottom: Scientific laboratories and a library for one million
books visited by the author in Lanchow, China. Boulevards
and buildings are being constructed in the midst of former fields.

Left: The television building in Peking, almost completed.

Bottom: Along the new railway to Turkestan and Sinkiang. Ancient trees and walls in the loess country of the upper Yellow River.

The caves of the Peking Man were discovered and partially excavated before war stopped the work. The remains of early man discovered are among the most important that have been made anywhere. No less than five skulls, besides many other bones—human and animal—and many primitive paleolithic implements were discovered. Primitive humans had lived in those caves on the banks of the river about a half a million years ago.

At that time the caves were close to the river, but by earthquakes and the inexorable processes of mountain building the hills had slowly risen while the river cut down its bed. Now the caves are on a hillside and from them one has a pleasant view over the plains towards Peking. At their foot the same river still wanders through the town. The Chinese, in their present mood of vibrant nationalism, like to think that they are the descendants of an indigenous line half a million years old.

During the Sino-Japanese war attempts were made to safeguard the five skulls. Unfortunately they disappeared. There was some muttering by my companions that they had been stolen by American imperialists and are today hidden in a New York museum, but no one was very explicit about this. It seemed that the fate of these bones is really unknown. Later a Canadian doctor who had spent his life in China told me that he believed that the skulls had been lost with some small ship during an endeavour to smuggle them out of China to a place of safety.

After a cup of tea, we were taken to the excavations, some of which we entered. In most places the original roofs of the caves had fallen down, so that most of the digging was now being done in great open pits. Sometimes the men blasted, but more often they carried away the boulders and baskets of earth. Paths and guard-rails had been laid out around the excavations, the whole area of which covered many acres. Lam Po-chia estimated that only one-third of the site had yet been dug.

"The present work was begun again in 1957," he said. "The work is under the direction of the Institute of Vertebrate Paleontology of the Academia Sinica, in Nanking. Most of the material has been taken there for study. Sixty men are employed and they have started the rough work of removing the fallen roof from another part of the cave."

We returned to the office for our picnic lunch and when we had finished our sandwiches, boiled eggs and tea we stretched out in arm-chairs about the room and slept. It was Sunday and rather hot.

On account of the blasting the excavations had been temporarily closed to the public. I felt particularly grateful therefore to the elderly scientist Dr. Lam Po-chia, who had made a special trip (on a holiday) to show us these workings, in spite of the fact that he looked tired and not in the best of health.

In the afternoon we walked farther up the hill to other caves, in which bones and implements of a second and much younger culture had been found, of the Neolithic Age. From the shaded hillside paths we had a splendid view over the fertile plains and the green hills that surround Peking in a great arc on the northern and western sides; and at our feet, to the south, were the workings and buildings of the new museum. Half a mile away on the top of a prominent hill I noticed the remains of a concrete pill-box.

"In the old days that was a lovely spot," said the old man. "The hills were wooded and on the top there was a temple; but in the war the Japanese burnt the forest and destroyed the shrine.

"It is not far from Peking. Many people will wish to come here and to see the excavations. It is intended that this should become the centre of a park of culture and recreation. For that reason grazing and quarrying are being stopped and trees planted. It is hoped we may restore these hills to their former beauty," and he pointed down the slope to the young trees and the abandoned quarries that were to form the new park.

My companions drew my attention to a cave across the river valley which they said was full of stalactites and stalagmites. It was being prepared for sightseers and they suggested that we should later visit it.

Down from the hill, by steep paths through sweet-scented trees, we strolled to the museum. It consisted of seven rooms on a single floor, in which were displayed some of the surviving bones of Peking Man, including five teeth that had recently been discovered, the relics of the neolithic culture of the upper caves, and many bones of animals that had shared the caves and fought for their possession with these ancient men. Life had not been easy. There were bones of queer extinct beasts like sabre-tooth tigers, giant bears, and strange forms of deer, with which Peking Man had struggled. From the state of the bones, he seemed to have eaten them more often than they had eaten him.

The specimens in the museum, with simple models and charts, made an appropriate and easily understood display for the benefit of the public. Although the museum was supposed to be closed, some people had managed to come in and they were showing a genuine interest in the exhibit. This interest impressed upon me the strength of the new and universal zest for literacy and learning in China. The exhibits interested me too, although all the signs were in Chinese except for the Latin names of a few fossils, and these did not mean much to me.

After another cup of tea we thanked our hosts, photographed them beside a pool of golden carp, and returned to our car down the garden path where Chinese were sitting relaxing in the shade. We drove across the sunshine of the hot valley to the other quarry.

This had by no means been opened to the public as yet, and the approach was up the dusty street of a little village. Quarrying had stopped, and when we reached the cave it was locked with a wooden door. We spent some time waiting and walking about the village looking for the local man who kept the key, but eventually gave up and drove back to the city.

I felt that the time was not wasted, for we had wandered from house to house around a village that was certainly not one normally displayed to the public. It very likely, therefore, represented an average state of affairs. It was untidy but peaceful, and every corner of land was cultivated. Children, pigs and chickens scrambled out of our way, and women watched us from doorways. As we passed, a man urged his horse, pulling a two-wheeled farm cart, closer to the walls that bounded the village street. The road was unpaved. It was muddy, rough and dirty. Perhaps the village was cleaner than formerly and organized as a commune, but the efficiency and bustle that led to the sweeping of every inch of Peking every morning had clearly not yet reached here.

Although it was Sunday, most of the men were using the fine summer weather to plant and irrigate the fields and most of the women were in their houses, busy with their children and their household duties. The children were running around in the sun and playing in the mud or dust outside the doorways. They had few clothes on and the boys under four wore none. They strutted about, proud of their bare brown bodies. The scene was peaceful, industrious, ancient and not at all miserable. The houses were old and the village had an air of accumulated untidyness. There was no outward sign of regimentation or evidence of any violent uprooting or upheaval.

On the way back to the city we passed a wooden triangular tower with a target on it that appeared to be a marker for geodetic surveys of the country. As I had seen these at intervals the whole way across Asia, and as Chow Kai was a geodetic surveyor, I asked him about this. I immediately regretted my question for he glanced at me suspiciously and virtually accused me of being a spy for American imperialists.

"All the maps of China were stolen by the Chiang Kai-shek clique," he said. "The American army has them."

I had not meant to upset him, and pointed out that as he must know from his visits, maps of the United States and Canada, on all scales, were freely sold to the public; therefore it had not occurred to me that I was asking an improper question. Gradually I succeeded in mollifying him, but it showed me how extremely sensitive the Chinese are to foreign intervention.

The Institute of Geology

Monday, September 1st

When I came downstairs at 8.45 A.M. Chow Kai was sitting in a stuffed chair in the lobby of the hotel. Mr. Tien came in and said:

"Good morning. I trust that you have slept well. Since we still have a little time before we need to leave for the institute, I hope you will excuse me for a few minutes, for there are one or two matters to which I should attend."

Indicating a chair for me, he bowed ever so slightly and disappeared up some stairs to a mysterious office on the mezzanine floor which he regularly visited but to which he never took me. That office seemed to guide my destiny like the unseen hand of fate. Half an hour later we arrived at the Institute of Geology of the Academia Sinica. It was housed in an old building, opening off a small concrete courtyard.

By this time I was aware that the Academy of China (to give it its English title) controlled about fifty such institutes, most of which do basic research in some field of science.

I was taken inside and greeted warmly in the director's office by the vice-director of the institute, Dr. Chang

Wen-yu. With him was another geologist, Professor C.-Y. Yang, a Ph.D. in paleontology from Yale University, who asked cordially after a number of Western paleontologists and told me that he continued to exchange type specimens of fossils with Europe and America, a matter of obvious scientific benefit to all concerned.

We sat down in easy chairs around a low table, but the cups of tea on it were soon displaced by geological maps which they brought out for my inspection. The first was a geological map of China on the scale of 1" = 60 miles, which had been published in 1948, and concerning which one of them remarked:

"This map was published just before the start of our new régime, but it is shortly to be replaced by a new and improved edition on the same scale that will show how great our progress has been. You will notice that there are many unmapped areas left blank on the old map, but during the last ten years most of these have been filled. The new edition will have blank areas only in parts of Tibet, which is very difficult to map."

It would have been easy to give political significance to this remark, but when I reflected upon the state of mapping of the mountains of Canada and of the United States, I realized that in Tibet the mountains were a sufficient obstacle and excuse in themselves. They also told me that good progress was being made in mapping all parts of China, except Tibet, on larger scales. They showed me copies of specialized maps shortly to be published on the same scale concerning the soils of China, its structural geology, and the areas that were subject to such disturbances as earthquakes, uplift and mountain building.

During this conversation I was considerably confused until they pointed out that there were two Institutes of Geology with identical names. This one in which I was sitting was part of the Academy of Sciences and the other was under the Ministry of Geology. The latter did straightforward geological mapping, while the one which I was visiting engaged in more specialized and fundamental research problems. This arrangement of the more practical institutes under ministries and of the more theoretical institutes under an academy is exactly that which prevails in the U.S.S.R.

We chatted until the ten-o'clock break for physical exercises was over. As usual, the senior men did not participate, nor would they listen to my suggestions that we go out and do exercises also. Then they led me to the lecture room, where a hundred or so young scientists had assembled. They all stood up when we entered, and after a brief introduction I gave my second talk, with the very able assistance of the Yale paleontologist.

Because of the extra time required for translation and for drawing diagrams, I did not finish speaking until nearly twelve o'clock. The senior scientists showed a keen interest and asked many questions until 12.30. Most of them had been at both talks, and they had had time to consider my earlier remarks. I was interested in finding how well they understood. Of course, the visit of any Western scientist was an unusual event for them.

Today, as before, the audience was most attentive and polite. The Communist influence has certainly not affected the good manners of the Chinese students or their respect for older or more senior scientists.

Question time was made more difficult by the tremendous racket of martial music that started outside in the courtyard at noon. It was evidently designed to raise the enthusiasm and nationalist fervour of the geologists while they were eating their lunch. At the time it was rather distracting; but to tell the truth, I enjoyed the gay airs with their Oriental flavour, although had I understood the words, I would probably have been very much annoyed. What I didn't know, didn't hurt me.

As we walked through the building, I was surprised to

see no wall posters. I have no idea whether this was because the staff were more politically astute or whether it was because the juniors of the establishment were less active. But in general, there were less wall posters in Peking than I was to find up-country.

As we came out of the building, I stopped to photograph some of the scientists on the steps. I noticed others drive off in a 1957 Belair Chevrolet, to which lace window curtains had been added. On the way back to the hotel in our Russian Pobeda, I noticed 1957 Fords and Plymouths, new Dodge trucks, older Chrysler and Cadillac cars, and a wide variety of English, German, French, Italian, Chinese, Russian and Czech buses, cars and trucks.

Dr. Chow Kai and Mr. Tien stayed for lunch. "Try this fried egg plant," they insisted, "you will find it delicious." Later they said, "If you haven't had consommé of three-delicious, you should try it. It's good."

I washed it all down with the inevitable tea and beer. Chinese beer is watery and about as bad as their tea is good. In the heat there were times when I longed for a glass of water. For three weeks I never touched a drop of it, and I never felt unwell.

After lunch I slept peacefully until three o'clock, when Mr. Tien woke me up to take me back to the institute.

"Please bring your passport," he said. "It is necessary to stamp an exit visa in it." Since I had crossed the Russian border no one had asked to see it, still less to take it away from me.

Back at the Institute of Geology, we went for a tour through many of its offices, libraries and laboratories. These were scattered in many rooms in several adjoining old buildings. Some of the rooms were empty or closed because in the summer the geologists were engaged in field work. Everything else was very crowded, but they said they expected to have a new building next year in the academy area northwest of Peking. They showed me many samples of minerals from newly discovered deposits. These included beryls and samarskites, useful in atomic research; celestite, which contains strontium; rock salt and manganese ores, ores of uranium and of rare earths. There were large feldspar crystals, of more scientific than economic interest, and large sheets of transparent white mica.

The last mica of such good quality that I had seen had been sent to me from Morocco two or three years before for the purpose of making age determinations. Some had been left over, and I remember persuading my brother-in-law, a New York brain surgeon, to perform the delicate operation of cutting and fitting mica windows into the doors of an old Franklin stove in our summer cottage. I had always felt a secret pride in having a stove whose windows had been so specially selected, meticulously fitted and accurately dated.

I asked the Chinese whether they had started to make age determinations by radioactive methods. "No," they said, "we have as yet no apparatus for determining the age of rocks in China, but we are considering the desirability of starting such work."

I also went through six well-equipped chemical laboratories. In one room special studies were being made of germanium, gallium and indium—rare metals useful in making transistors. In another laboratory they were studying the rare earth elements, and in another analyzing the ordinary elements of common rocks. I saw several spectrographs and some x-ray equipment for studying minerals made by Shimadzu in Japan. Elsewhere I saw microscopes, polarimeters, flame photometers made by Zeiss in East Germany, and other instruments from Czechoslovakia and Russia. Most of the simpler equipment had been made in China and all the chemicals had Chinese labels. One Geiger counter had been made in California, but I do not know at what date.

They were apologetic about their library, which was

small and deficient in older books. They complained that with the growth of the number of institutes, others had obtained most of these; but they were subscribing to many modern scientific periodicals and were anxious to order more Western technical books. They asked me for the names of recent works on geology and geophysics. Although I mentioned a considerable number, I was able to suggest only two of which they had not yet heard. One of them had been published in 1958 in Berlin. It was quite evident that many of the Chinese scientists not only spoke Western languages, but also read a great deal of recent geological literature from all over the world.

They gave me a large number of reprints of their own papers, written in various languages, to which were generally appended two abstracts so chosen that the main work and the abstracts were in Chinese, Russian, and some one of the principal Western languages.

One of the few research men who was not in the field showed me his study of Chinese phosphate deposits, which are important as the basis of valuable fertilizers. He seemed to be making a thorough and modern study of the problem. On maps he pointed out the localities of phosphate-bearing rocks, of which different types occurred in the north and south of China. He had more detailed maps of separate occurrences which showed their thickness, type, and the percentage of phosphate that they contained. In the most favourable places the rocks contained 38 per cent of phosphate, which made them rich fertilizers.

I got the general impression that the institute was devoted to the solution of practical problems and that these geologists were tackling them by sound and up-to-date scientific methods.

Mr. Tien drove me to the British Embassy, where he left me and where one of the very affable secretaries and his wife gave me tea. Though some of the bachelors came to the Peking Hotel for a meal, on the whole the embassy people were more isolated from the Chinese than were visitors.

The outside walls of the embassy were still hung with the tattered remnants of the anti-imperalist signs. It must have been nerve-wracking to be besieged by a mob for thirty-six hours and to be the object of a virulent hate campaign, but they preserved remarkable sang-froid. They agreed with my views on the excellence of Chinese workmanship and with my suggestion that there was every evidence that the present régime would be permanent. I mentioned the Quemoy shelling, which was then at its height according to the news they quoted from the BBC. They agreed how unfortunate it was, but doubted if it would lead to serious consequences.

On parting they said, "On your way back to the hotel you should visit the Marco Polo antique shop. It has excellent things, and it is not far out of your way. Turn left as you leave the embassy gate. Walk two blocks past the Russian embassy, which you will be able to recognize by the number of cars in front of it, and turn right. The shop is across the road in the middle of the first block."

I spent the evening in my hotel room and this gave me time to consider the Peking Hotel. A brochure that I had discovered in the desk stated that "electric fans, flower pots and radio sets are on hire," and that "each floor is provided with electric irons and refrigerators for the benefit of guests." The brochure concluded with the polite remark, "Criticism and guidance relating to the inefficiency, inadequacy in our service are cordially welcomed with thanks."

The hotel was all very clean. The hardwood floors in the corridors and lobbies were polished every morning. The servants were all polite and clad in Western clothes. The men wore white coats and dark trousers; the girl waitresses were uniformly dressed in heavy white silk blouses, black skirts with white aprons, white short socks, and black shoes. They were pleasant, efficient girls who quickly got to know my tastes and seemed happy to fill my teacup innumerable times.

Most of the servants were quick and alert, but one of the waitresses in the upstairs European dining-room had slowed down a good deal and sometimes sat down for a rest. When she walked she leaned very far back on her heels for rather obvious reasons.

I was particularly charmed by the chief waiter in the Chinese-European restaurant downstairs, who was a most gay and efficient gentleman with a good command of English and a ready wit. He gave one the impression of having a real enthusiasm for his work of looking after the varied assortment of guests from many countries who frequented his dining-room.

A curious feature of this hotel that stood out in contrast to the hotels in every other part of China was the emptiness of the public rooms. They seemed to be reserved for government functions. The only one I saw was a cocktail party on the open roof garden one evening. I believe that the hotel is the largest in Peking and reserved for visitors from Western and neutral countries, most of whom were there as guests of the Government. No Russians or people from other satellite nations, no salesmen, no business men ever came to this hotel. Evidently, they had their own. Chinese never came alone but only when accompanying guests from abroad.

THIRTEEN

Passports and prospecting

Tuesday, September 2nd

I was up at 7.30 and went to the European dining-room upstairs for breakfast of an omelette and tea. I felt that I was not cheating too badly in my resolve to eat only Chinese food, for I had learned by now that black eggs, sour pickles and stewed rice were not the only form of Chinese breakfast. I could have had an omelette and tea downstairs. The wait-ress was rather astonished when I refused a knife and fork and asked for chopsticks.

Going down to my room I noticed that the elevator was of a make well known in North America, but it had on it in English "Made in Switzerland." It worked well, either with an operator or in the evening by automatic buttons.

"*Wu lou,*" I said to the elevator operator. Whether he understood that I was trying to say fifth floor, or whether he merely recognized me and knew where I should get off, I do not know.

In my room I was preparing to go to the University of Peking when Mr. Tien phoned and said that there had been a change in plans. I wondered whether this delay was be-cause of some real difficulty or whether I had asked to see too many institutes. In any case I was glad of the chance to write letters and to get my luggage organized. It was nearly noon before Mr. Tien arrived. He was rather brusque and asked me for two passport pictures.

"It will be necessary for you to get a special police permit to stay in this country."

"That is absurd," I said. "I have been here a week already and no one ever mentioned the matter before. This is a ridiculous piece of red tape."

I was puzzled, but the reason soon became apparent when Mr. Tien produced my passport and said,

"Why does your passport refer to 'Mainland' China? There is no 'Mainland' China. There is only one China, not two Chinas. It is most insulting."

I asked to see the passport and discovered that he was referring to a slip of paper, giving instructions to Canadians intending to visit Communist countries. It had been pasted in the back of my passport by the Canadian authorities when I had asked their advice and help in getting visas for this trip. The notice warned Canadian travellers on entering or leaving Communist countries to report to Canadian consulates on both sides of the border, for their own protection. If there was none, Canadians were advised to go to British consulates; and it gave a list of these countries and addresses of the consulates.

No doubt with the best of intentions, and because Canada does not formally recognize their governments, it referred to two of these countries as "Mainland" China and "East" Germany, instead of by their proper names, "The People's Republic of China" and the "German Democratic Republic." The whole thing seemed unnecessary and a great nuisance.

Fortunately, I realized that the offending paper was not part of the passport but merely a note of advice. I therefore ripped it out and tore it to small pieces.

"Stop!" said Mr. Tien. "You can't do that. You are defacing your passport."

129

"Nonsense," I said. "It is not part of the passport, and I've done it." I gave him back the passport and, muttering, he went off to the authorities on the mezzanine floor. I realized that I had caused the Canadian Government and myself a considerable loss of face, but it was a small price to pay for the resultant restoration of my passport and Mr. Tien's equanimity. Thereafter, all question of an internal passport was dropped.

Peace restored, Mr. Tien and I set off to cash some traveller's cheques at a nearby branch of the Bank of Peking. The building was an ordinary bank and had undoubtedly been there for a long time. The teller accepted my American traveller's cheques without surprise, giving me in receipt a numbered metal disk like a cloak-room tag.

He made some calculations on an abacus, entered them on a form, and counted out a wad of bills in small denominations. (The largest bill I saw in China was for ten Chinese dollars which is equivalent to four dollars in Canadian money.) These he passed to a colleague who checked the calculations, recounted the money, and stamped the paper in red ink with his little ivory "chop" or personal seal.

The first clerk then came to the counter, called my number, took back the metal tag and paid me the money.

Pocketing my roll, I got into the car and we drove to the old and new markets and out to Embroidery Street, to make those purchases upon which I had decided as a result of my previous visits.

Back at the hotel with my presents, I spread them out on the bed and gloated over them. Compared with the amounts in the shops, my little collection was small, but it pleased me. In some countries there are few good things to buy, in others they cost as much as at home, but here I had gifts both cheap and beautiful.

I sorted out some of my warmer clothes to mail home and considered the purchase of a string bag to carry the still remaining overflow to Japan, where I could mail home my tropical clothes as well.

After lunch two boys from one of the shops came to my room for a second time, loaded with boxes of jewellery. Their eloquent assaults on my purse were useless, because I had already made my choice, but this survival of high-pressure selling in a Communist state interested me.

At half-past two Mr. Tien reappeared and announced that the plans had been adjusted. We would go to the University of Peking tomorrow, but we must now go to the Institute of Prospecting. The driver took a new route, and once outside the city walls he and Mr. Tien became completely lost.

We cruised around aimlessly in the rain through a series of new and muddy boulevards and streets, surrounded by unending stretches of grey academic buildings in various stages of construction. This meandering gave me a further opportunity to gauge the area of these institutes; but plans had twice gone wrong and this gave poor Mr. Tien a serious pain in his *amour-propre*. Any failure of a plan is taken very much more seriously in a Communist country than it is with us.

Finally, at 3.30 P.M., we arrived at the Ministry of Geology's Institute of Geological Prospecting. We were greeted in the Administration Building by the vice-president, Professor C.-T. Chung; the professor of mineralogy, Dr. P. Shu; and our old friend Dr. Chow Kai, the professor of geophysics there.

While we sat in comfortable chairs and enjoyed a cup of tea, I was told that the institute had been founded in 1952 and consisted of six departments: geological surveying, geological prospecting, geophysical prospecting, mining engineering, petroleum engineering, and hydraulic and engineering geology. They said that of the five hundred teaching staff, only about thirty were experienced professors like themselves, but that with the aid of junior assistants they managed to cope with six thousand students. Until 1954 they tried to train men in four years, but now the course was five years long. At the time I was there most of the students were away on prospecting parties, of which they had this year sent seventeen to various parts of the country.

The demand of the Government for prospectors seemed insatiable, and they were still expanding, so that the enrollment might reach 7,500 this year. In addition to this, they gave short courses and had night-school and correspondence classes, for they pointed out that they had to train technicians as well as engineers and scientists. So far they had graduated only two thousand students from four-year courses and three groups from a shorter two-year course.

In spite of the tremendous amount to be done by the limited number of senior staff, they claimed to have started some scientific research two years ago and to have published two works already.

They had organized a museum, which they showed me. It was arranged in about twenty small rooms, containing altogether, I estimated, about seven thousand rock specimens. Most of the exhibits dealt with practical matters such as coals, metallic ores, non-metallic deposits, or petroleum deposits. They also had many specimens of rocks that are associated with different types of mineral deposits and might give a clue to their discovery.

Other rooms were of a more scientific nature, containing systematic collections of minerals, rocks, fossils, or geological maps. A few rooms dealt with miscellaneous items such as Chinese meteorites, the local geology around Peking, and the model of the first seismograph—which, they told me, had been invented in 132 A.D. by a writer and scientist named Chung Hung.

Some of the mineral specimens were left over from old collections. For example, many of the crystal models and rock specimens were familiar to me and had evidently been brought from German supply houses many years before, but most of the mineral specimens appeared to have been collected recently in China, and there were many boxes and piles of material that had just arrived from the field.

There were signs of Russian influence in the form of maps and in a series of photographs. In the room where we had tea I noticed portraits of Mao, Chou En-lai, Stalin, Lenin and Marx. Other rooms had pictures of Russian scientists and of a very few scientists from other countries.

As we left the Administration Building, I was told that it had been completed four years before and was the oldest building on the campus. It was a large brick structure, four stories high.

We next examined the machine shops. These were mostly small brick buildings of very simple construction, which they said were temporary. The fact that temporary buildings were always of matting or light brick illustrates the desperate shortage of wood in China.

In one of these huts were eight Chinese lathes, a milling machine, and a planer, operated by some thirty students. In other buildings there were a forge and furnace, a mechanical hammer, a compressor, and a foundry where they were pouring castings into earth or clay moulds. What surprised me was not that it was all rather simple and primitive, but that it was there at all and that it all worked! They showed me a drilling machine and pile driver that they were constructing, even to the extent of casting the pulleys.

It was raining and the paths were exceedingly muddy. Sunflowers, castor-oil plants, corn and cabbages were still growing among the huts, and without any doubt the whole area had been fields until recently.

The next building was a large, four-story chemistry building, full of well-equipped laboratories. In the balance room I noticed Japanese and Chinese instruments. In another room I was shown a spectrograph, which they told me was being used continuously, two shifts a day, for practical work. I saw a Hungarian polariscope, but they explained that a better one with automatic recording was now being made in China. In this building we came across a group of a few-score students who were being trained to make analyses by rather primitive methods. They told me that these were technicians taking a three-week course, during which they learned ten specific tests in soil analysis.

The next large, modern building contained empty lecture rooms and laboratories devoted to geophysical prospecting. The electrical, seismic and magnetic equipment was quite conventional, but I was surprised at the abundance of it. Outside the building was a six-wheel-drive truck, equipped for electrical prospecting.

A sketch of the campus which I made on the spot shows the position of six major class-room and teaching buildings, of which I have just mentioned three, two large dormitories, and two more large buildings under construction, besides temporary huts. There were two heating plants and a sewage plant in sight, and they pointed out that the adjoining institutes were concerned with aeronautics, with medicine, and with mining. The head-frame of a practice mine shaft towered above the latter. They said that they regretted they had as yet no separate library building but hoped that it would be built shortly.

They emphasized that in teaching they combine theory with practice, and that all students and most professors spend all summer on prospecting surveys, which explained why there were so few students about.

I saw basketball courts and gymnastic equipment. There were loud-speakers hanging about the place, blaring propaganda and martial music. After the Russian fashion, there were several life-size plaster statues about. They had been designed for propaganda purposes rather than as art, for they portrayed buxom lads and lasses joyously engaged in useful work or games. At the entrance to the Administration Building was a huge coloured poster. Against a background of mountains a girl holding a geological hammer was pointing the way to her male companion. What would Confucius have thought?

FOURTEEN

The University of Peking

Wednesday, September 3rd

At breakfast I shared a table with Norman Endicott, one of the party of five Canadian lawyers. His father, after whom he is named, had been a missionary in China, and the son had been born and spent his first fifteen years in Szechwan Province. He spoke the Chinese dialect of that district fluently and understood Mandarin.

"Today in China," he said, "there is no private practice of law. All lawyers are employees of the courts or of branches of government. There is much debate about legal matters, but the conclusions always seem to follow the party line as expounded by Peking. The only Chinese who feel at all free to express independent opinions are those veteran Communists who took part in the Long March to Yenan in the nineteen-twenties."

Half an hour later I saw the full beauty of Peking as Mr. Tien and I drove past the gate of the ancient Forbidden City. The white marble steps with their carved balustrades shone brilliantly in the early sunlight against the dark crimson walls of the vast fortress. The great new square gave one a splendid view of the ancient palace, for this was an ordinary work day and it was empty except for the swirl

135

of traffic along the main east-west road across it, and the eddy that flowed off towards the railway station.

The traffic of vehicles and bicycles, of hand barrows and horse-drawn carts, bobbed along like flotsam. Trucks and buses darted along the spacious pavements. Ancient brown-and-cream street-cars crowded with people rattled along over their narrow-gauge tracks. Little schools of bicycles and pedicabs sprang to life as the traffic lights changed at intersections. Many of the bicycles were polished and shone like new American cars; but, as Mr. Tien remarked:

"All the pedicabs are old, for they are not being replaced. It is not fitting for men to act as beasts of burden."

Between the shafts of the two-wheeled Chinese carts thin horses waited patiently while the white-coated traffic police-man waved us through. The fact that every cart had rubber pneumatic tires was a sign of progress, but the canvas bags between the shafts behind each horse, to catch droppings, was a sign that the Chinese had not lost their frugality.

Passing along the boulevards through the new academy district, we came to the University of Peking. It is clearly an old campus, formerly known as the missionary Yenching University, I believe. In front of the Administration Building we got out of the car and stood around waiting until we were joined by Dr. T.-F. Hou, the director of the department of geology and geography, Professor S.-T.-C. Wang, the professor of geophysics, and Miss Fu, an interpreter. Through wooded gardens of the old campus we walked to a pleasant guest house, where we had a long discussion about Chinese education.

"In primary school," said Dr. Hou, "most students are now being taught Mandarin, which is the dialect of the Peking district. It is readily understood in most parts of north China and is comprehensible to the people of Szechwan in the west, but in the south the Chinese dialect languages have become greatly modified and Mandarin cannot be understood.

"This modification of the language is similar to the changes that have occurred in European languages derived from Latin. In effect this will mean that many of the children in China will become bilingual, for their native dialects will no doubt be preserved in the home but the government will insist that everyone learn Mandarin at school. There are two purposes in this: one is for efficiency and convenience in communicating the spoken work, whether in business or by radio, and the other is in preparation for the alphabetization that is now being introduced.

"It has now been agreed that over the next ten years the Roman alphabet will be introduced, and a new system has been devised for the phonetic spelling of the Mandarin dialect. In future everyone will be taught to write Chinese in Roman letters."

This system is, of course, similar to the way in which we now transliterate Chinese names like Peking or Shanghai into English. Although from pride they always said Roman, it is in fact the English alphabet of twenty-six letters that is to be used. Simple as it seems to us, the change is a tremendous and significant one for China, and it is interesting that it is to the English and not to the Russian alphabet that the change is being made.

The change has been forced on the Chinese because reading, writing, printing, typewriting and, in particular, the making of carbon copies are all so much easier with an alphabet of few characters instead of with thousands of ideographs. It was of necessity and for progress that such a profound change was being made. To abandon the ideographs that the Chinese have used for thousands of years was a hard decision for so proud a people.

"Among the 650 million people who live in China," said one of the professors, "there are between fifty and a hundred million who belong to national minorities. These are the people like Tibetans, Mongols and Manchus, who are not of the Han race and who have languages of their own quite different from Chinese.

"These people are to be allowed to study their own spoken languages in primary schools. Fifty-one languages of the minority groups are now officially recognized. They are all being standardized. Dictionaries and grammars are being prepared. In some cases, it has even been necessary to devise written forms of the languages when none existed before. These people will be allowed to have newspapers and books in their own languages.

"All primary schools are free and everyone is being taught to read and write. In every town secondary schools are being established. There students start to learn a Western language, generally English, Russian or German. There are not yet enough high schools or enough high-school graduates to fill all the places available at the universities. Although we are building new secondary schools the demand for secondary-school matriculants is far greater than the supply.

"At college the students have to learn one foreign language so well that they can read, write and talk it fluently. They generally start a second language which they are expected to be able to read, but graduate students have to know at least two foreign languages fluently. These may be any of English, Russian, German, French, or Japanese.

"At college all students get free tuition, free medical services, and a place in a dormitory free. They have to buy their own books, clothes and food, but many who cannot afford to do this are provided with scholarships.

"Whereas the hours of class work were formerly more than thirty a week, this year classes are being reduced to twenty-six hours, which will for scientists include instruction in languages, physical education, mathematics, and political subjects, as well as science. At the same time the course leading to a diploma [the equivalent of a Bachelor's degree] is being extended from four to five years, so there will be no graduating class in 1959.

"The freshmen receive 180 hours of political indoctrination a year," (I rapidly calculated that this comes to about one hour a day) "but in the senior years they receive only two hours a week."

There are now in China, they said, two hundred colleges and universities, with an enrolment of approximately 450,-000 college students. About a third of the students are in Shanghai and about the same number in Peking; the rest are scattered. In Peking there are now more than thirty colleges, universities and teaching institutes.

It had been the intention, they said, to group them all in the north-west part of the city and many had been built there, but plans had recently been changed. In future new institutions will be scattered about the city.

Among the institutes they mentioned besides the University of Peking were the Peking Teachers' college and the Peking People's University, each with more than ten thousand students. The latter was still located, and perhaps would remain, in the centre of the city. Near Peking University was the campus of Tsing Hua University, which was an engineering institute with more than ten thousand students.

In another building they showed me a model of the campus of Peking University on which the buildings had been painted different colours to show whether they were completed, under construction, or only planned. I counted about a hundred large buildings, of which five were under construction, including a large new library building, and twenty-nine were planned. Of the remainder I estimated from their appearance that about half had belonged to the old campus and about half had been built since the change in régime.

A feature of the campus that they pointed out to me with great pride was a row of very small houses, which they said were individual homes for senior professors. So far from boasting about communes, they regarded the idea of individual family houses with great approval.

Later we walked around the campus, and as far as I could

recollect the model was accurate and up to date. The older part of the campus had Ming-style buildings on shaded walks and was among the loveliest campuses I have ever seen. It was one of a dozen missionary universities. I was shown the old library building. They were extremely proud of the fact that it contained two million volumes, of which one quarter were in foreign languages. The old buildings had not been changed noticeably and appeared graceful and beautiful in their shaded campus beside a lake. I suddenly exclaimed upon the beauty of a pagoda and its reflection in the water. "That," said a professor with a smile, "is the water tower."

Most foreign students coming to China study in Peking, and at the present time (1958) there are about three hundred foreign students here from twenty-four countries. The University of Peking also makes a speciality of teaching foreign languages, and for that reason there are perhaps a dozen professors who are foreigners, teaching their own languages such as Indonesian, Arabic, or some of the Indian languages.

I heard later that there is still one English professor at this University, who has a Canadian wife, but I did not meet them. They both teach English, and recently visited Toronto and England on furlough. They used to teach in mission schools.

We then discussed the work in geology and geophysics and visited these laboratories. They were in the newer part of the campus. "Formerly," said Dr. Professor Hou, "this university had a large department of geology but it was split off a few years ago to form the Institute of Geological and Geophysical Prospecting."

"The institute that we saw yesterday," interjected Mr. Tien.

"In 1955," continued Professor Hou, "a new department of geology and geography was started, but unfortunately on this occasion all the faculty and students except two

140

professors and a few graduate students are in the field, engaged in practical work for the Government. In field parties of staff and students together they have fanned out over our great country, all the way from the northern frontiers along the Amur River to the south coast of China, and from Shantung Peninsula to the upper reaches of the Yellow River."

He said that because of the importance of this field work the department of geography and geology had a summer break that lasted one month longer than that of the rest of the university, which normally lasted from the first of June to the first of September, and that whereas they had four hundred students last year, this year they expected a total of six hundred. The teaching staff, which had been forty last year, would also be increased.

He admitted that the teaching program was being reformed at the same time that the courses were being lengthened, and that they did not yet know exactly what the new teaching program would be.

"All the students are in residence," he said, and at one place we passed a cafeteria where students were eating rice from bowls held up to their mouths.

I gathered that the students had no more space than necessary for sleeping and sitting at a desk. I also understood that the average faculty family lived in dormitories on the campus and had a total of ninety square metres' floor space. That would be about the equivalent of three rooms each fifteen by twenty feet.

Professor Hou was a geographer, and he explained that the geography section was chiefly concerned with planning and with studies of natural resources and land use, but that he himself was a specialist in historical geography for the advanced course in which students had to learn ancient Chinese—the equivalent of our Western classical languages. At that time geography and geology were sharing a building with chemistry, but during the winter they expected to get a new building.

Physics was housed in another of the old buildings. In the departmental library I noticed about 250 periodicals, of which some fifty were in Chinese and fifty in Russian, and the rest in various Western languages. I saw copies of the English journal *Nature* with back numbers to 1890, and of the American publication *Physical Review*, and of the *Canadian Journal of Physics*. The latest copies of all publications I saw were, with one exception, dated 1958. The idea never occured to me in China, but I have since wondered if there is an embargo on the export of American journals to China. Certainly there was no sign that it was effective.

Although the physics building was old, it seemed to have rather more and better equipment than at some other universities. There were a few students in the laboratories and more seemed to be busy in political matters, for the walls were covered with posters and on many of the blackboards were chalk sketches showing the natural artistic ability of some of the students. Some of these celebrated the *sputniks*, and others the abundance of the 1958 harvest. This was a propaganda line that I saw on posters elsewhere.

Geophysics was housed in two huts adjacent to the main physics building, and a group of students there were building some electrical prospecting equipment. When our party came in they stopped work and smiled and bowed and clapped their hands. We smiled and bowed and clapped back. The same performance was repeated when we passed through the sports area. In a modern gymnasium a group of both men and women were engaged in gymnastics and tumbling. They clapped when we came in and then for five minutes staged a very quick and excellent exhibition. When it was over it was our turn to clap, but they joined in again. On a track outside men and women were practising running, at the same time but in different groups and not in competition with each other.

After leaving the university, we drove past the government offices and dormitories along the west side of the city and passed the new television building, a fine ten-story affair, with a high television mast projecting above a tower on the roof. Although it appeared to be practically finished I was told that the building was not yet occupied, but that preliminary experimental broadcasting had started for two or three hours a day in the Peking area and that a proper program would be launched in a few months' time.

The faithful Mr. Tien and I were joined at lunch by S. P. Lee and Chow Kai, who now seemed like good old friends, and with them we had an excellent farewell lunch with many toasts together in wine and beer. They gave me a farewell present of two lengths of silk that I had admired but not purchased.

Mr. Tien asked for my films. He said, "You will have to give them to me to be developed, and I will return you the pictures when we get to Canton."

"But," I said, "I don't think that you can process colour films in China."

"I am sure that it is possible, and I regret that I will have to ask you for four yuan ($1.60) for each roll."

"But the cost of developing the film was included in the price when I bought the film in Canada."

"I am sorry, but we have no means of collecting that."

Dolefully I handed over the films and the cash, and we drove to the Summer Palace, where Mr. Tien purchased tickets for four cents each. "Children," he said, "are admitted free." Along with many Chinese, we entered and walked through the lovely gardens that had been built at the end of the last century by the Dowager Empress.

Since she did not think the flat plain of Peking sufficiently beautiful, she had employed tens of thousands of men with baskets to excavate a large lake. It must extend over many tens of acres. The soil had been piled up to form a large hill. Upon one side of this she had built a pagoda in a rather decadent Ming style of architecture, and other palaces and theatres for herself.

143

The whole place was open to the public and a considerable number of Chinese were strolling about. Some were out in boats fishing on the lake, and a great number of small boats were for hire. On week-ends thousands come to the park to enjoy themselves.

We visited the beautiful quarters in which the Dowager Empress had for a long time imprisoned her nephew, the rightful Emperor. Although it had every appearance of being a palace, it was in fact a prison in which she kept him, while with a fine defiance of all Chinese custom, she retained the throne that was rightfully his.

We saw the triple theatre with which she indulged herself. In it there were three stages set one above the other, on one of which she sometimes acted, while at other times three performances were presented simultaneously, one above the other, while the Dowager sat looking at whichever show pleased her most. We went on to the marble boat, a tea-house beside the lake, which she is said to have built to commemorate the fact that she used money voted for the navy to build the lake.

In great haste we hurried back to the hotel for a quick supper and word of good-bye. I briefly saw two of the Canadian lawyers, Norman Endicott and E. B. Jolliffe, who were elated at having returned from an interview with Premier Chou En-lai. I wished that I could stay to hear about it, as I never met any political figures, but Mr. Tien rushed me to the station.

It was an exceedingly warm evening, and we had done well to visit the university and the Summer Palace and get packed by five o'clock.

Fortunately, the station was only five minutes from the hotel. In a welter of taxis we seized our bags and struggled through the swarm of waiting Chinese. They crowded every doorway, gate and corridor and with their children and bundles slept on the benches and overflowed onto the floor in the stifling waiting rooms. What a relief it was to squeeze past the ticket collector and emerge from the crowds into the comparative quiet of the platform beside our express!

It was still a long walk up the fourteen-car train to Sleeper No. 5, the only "soft" pullman on the train. Suddenly I realized that my bag was very heavy, that I was very hot, very tired and wringing wet, and I had only five minutes left to say good-bye to my good friends the Chinese geophysicists and to those officials who had so politely come to see us off.

After the whistle and the final handshake, the scramble on board and the last wave, Mr. Tien and I sank back in our seats to catch our breath. Until that moment I had managed to restrain the tremendous curiosity within me. As casually as I could, I asked,

"Mr. Tien, where is this train going?"

"It is the Chungking Express," he said, with a smile of satisfaction for what he knew to be my triumph. We were going up the Yellow River!

I looked around the car. It was plain and reasonably comfortable. In each compartment were four upholstered berths, or seats, on each of which was piled a thick blanket, two sheets and a pillow. Beside the door sat our only companion, a smiling but completely silent Chinese in a blue cotton uniform. His dress was typical of most Chinese — a blue peaked cap, a blue cotton jacket with a turned-down collar buttoned to the neck, and blue jeans. Below his blue jeans one could see his black cotton slippers with white rope soles. They had only a narrow opening for the foot, so that no laces were necessary.

Mr. Tien sat opposite to me beside the window. Between us was a little table with a lamp on it. Perhaps as a concession to his work with Westerners, he was considerably better dressed than the average Chinese. He never wore a hat but his suit was well cut from heavy, buff-coloured gabardine. It was a copy of the costume always worn by Chairman Mao. I could not help wondering whether is was indigenous to China or whether it had been copied from Stalin.

145

A month before, in the Caucasus, I had been struck by the fact that all the Georgian men, unlike the Russians, wore jackets of similar style, with turned-down collars, cut from white or drab material. Somehow I always supposed that Stalin's and Mao's dress had been adopted from that of workers, but it occurred to me that perhaps this was not the case — that Stalin was merely wearing the standard costume of his native Georgia. It was strange to think that perhaps the Chinese had thus copied the dress of a distant mountain republic.

Mr. Tien had a square, flat face with massive features beneath an unruly shock of almost crew-cut hair, which was black except for an occasional fleck of grey. He had seen hardship, but it had not hardened him. He knew English well but knew little about the West, and most of what he thought he knew about it was quite wrong and distorted. Since he was keen and intelligent, it is possible that my vigorous defence of Western capitalism may have shaken him a little, but I doubt it, for his indoctrination had been strong. However, our differences never led to bad feeling, and he looked after me exceedingly well. At that moment he was smiling happily, for he obviously preferred a trip to Lanchow and Hong Kong to his usual work of translating technical journals from English into Chinese.

Soon after the train pulled out, the voice of a young woman, with which we were to become very familiar, began a running commentary over the loudspeaker. She interspersed her chat with gramophone records and selections from radio broadcasts. Mr. Tien said that she was giving news, weather reports and accounts of the history and the future (bright I'm sure) of all the places we passed. Her talks I could not understand and hardly noticed, but the music was all stimulating and I enjoyed it. The only thing to which I really objected were the broadcasts by Chinese clowns. They are very skilful at making the most extraordinary sounds, which I found to be a distracting form of mental torture.

146

Soon an alert young porter, also in blue denims with a red numbered badge, appeared and put four large tea mugs with lids on the table. For four cents each he filled the mugs, and also a big thermos bottle under the table, with nearly boiling water, and gave us each ten tea-bags. On to the knob of each mug he placed a rice-paper receipt so marked with ideographs that each could recognize his own pot. Every hour or two he came around and filled up the mugs, so that for this tiny sum we were assured of all the tea we could drink for the next twenty-eight hours, and that was a great deal indeed. After one stop, where we walked up and down the platform in the dusk, the radio ceased, the lights were put out and we lay down on our berths.

I slept only fitfully, but it was not for lack of comfort. Through the window, which was screened and open, blew a gale of cool night air. Contrary to what I had always imagined, the night air of country China was sweet.

FIFTEEN

Sian train

Thursday, September 4th

At six o'clock we were awakened by the first newscast of the day. Gay morning music followed, which consisted chiefly of girls singing and of martial music. One girl sang a love song punctuated with much banging of gongs, another a plaintiff mountain air that Mr. Tien said was Tibetan. Then a girls' choir chanted a soaring hymn that I might have taken to be religious, had I not supposed that it dealt with Communist heaven. Best of all I enjoyed the marches and the working songs, which were like Western military tunes with an Oriental flavour. Mr. Tien identified one, saying, "That is the song of the liberation army," and another, "That is the hymn of the Communist Party."

We went into breakfast in the crowded but efficient diner, which was the next car. It was simply furnished for forty-eight people and lined with plywood, upon which the only decorations were lamp brackets and coat hooks. Each of the latter supported a neatly crossed pair of fly swatters, which were undisturbed because no one had any coats to hang and there were no flies to swat. It was scarcely spotless, but a waiter in a dirty white coat wiped off the oilcloth in front of me and put down a metal spoon and two chopsticks.

Since I was the only Westerner on the train, he had a

consultation with Mr. Tien. I interrupted to tell Mr. Tien that I wanted whatever he and the other Chinese had, and so for breakfast the waiter brought me a bowl of rice, some pieces of chicken and green pepper in sauce, and a gelatinous broth. I refused beer. It was certainly a Chinese breakfast, but Mr. Tien and most of the other passengers had simpler food, such as rolls of steamed bread with bowls full of noodles and gravy, or rice and fried vegetables.

The waiters were efficient and their chief problem in keeping the place tidy (which they tried to do) was with the passengers, who were a cross-section of the Chinese people, mostly peasants and nearly all young. They held their bowls of noodles up to their mouths and sucked the contents in with a slurping sound, scattered their rice on the tables with their flying chopsticks, and used their fingers to pick pieces of meat out of the gravy. They chewed the bones and threw them on the table or the floor .About twice every meal Mr. Tien used to leave the dining car in order to spit, clearing his throat ominously as he departed. Other passengers were less particular.

At first this tended to diminish my appetite, but one becomes accustomed and I was hungry from having slept with my head in a cool gale by the window. Everything was new and fascinating and the Chinese food was, as always, delicious.

Back in the sleeper the porter produced a can of hot water, and after rinsing out a wash-basin filled it for me. The facilities were simple but adequate, and it was typical of the effort being made to keep things clean that the porter also mopped down and dusted the whole car very thoroughly twice during the day. At intervals between filling our pots with tea, and especially after we had stopped at a station, he would sweep the car and go around with a fly swatter, but the only fly I saw him chasing escaped into the next car.

"They have a competition every day," said Mr. Tien, "and the porter with the cleanest car gets a flag."

I forgot to ask him who was the judge. The only likely person seemed to be the security policeman who occasionally wandered through the train. In the porter's compartment I saw him checking the tickets and passes of the passengers, but he never bothered Mr. Tien or myself and no one asked to see my passport.

From a station platform we identified the young woman in the baggage car who operated the loudspeakers. Shortly afterwards I noticed a change from the Oriental love songs, Western band marches, and girl choirs to a male voice that reminded me of Paul Robeson. He sang several songs interspersed with excited comment from the young lady. One of those songs was repeated, and Mr. Tien said, "That is the song of the Yellow River, which we are approaching. It has been China's sorrow, but it is now being dammed and brought under control and will become China's joy." We were running down a slight slope in the vast plain that we had been crossing all the way from Peking. As we reached the great river Mr. Tien continued, "The Government is now organizing the workers."

This was indeed true. As the river came into sight I saw an army of blue-clad men and women on the banks. They were building a railway embankment and river levees, and they were operating a factory for manufacturing piers. Two hundred yards down-stream from our track, another parallel bridge was under construction so that the railway could be double-tracked.

If not the widest, it is the longest river in China, and one of the great rivers of the world .It was formidable enough. Although it was low water, the river where we crossed it swirled through the trestles in great turbulent floods separated by islands of mud. It eddied under the bridge in whirlpools as thick and opaque as cocoa, its colour a deep tan from the enormous quantities of mud and loess it carried. On both shores, besides factories for making concrete, there were masses of stores and equipment of all kinds, including what appeared to be the remnants of a pontoon bridging outfit such as army engineers use. It was khaki-coloured and looked like American equipment that had been left behind by Chiang Kai-shek's army.

After crossing this formidable barrier, which had for so long divided China, we passed several large irrigation canals. Mr. Tien told me of the dams they were building to control the river, but none was in sight.

The land over which we travelled south from Peking was old and densely populated, part of the plain of eastern China where every inch is cultivated like a market garden. The price for supporting so dense a population has been to crowd the people into tiny thatched and tiled villages, which rise like islands in a flood of green extending unbroken to the horizon. Only a few narrow ditches and tracks break the fields, and a few remnants of old mud walls that had once divided them into smaller fields.

At intervals there were wells that thin and scrawny horses endlessly circled to drive the pumps. Here and there groups of a hundred or so people weeded and cultivated, or drove pairs of oxen pulling primitive wooden ploughs through the soft alluvium of transported earth. They had no conspicuous overseeers.

To this ancient land a new vigour is being imparted. At most towns there is a brand-new stone station, and on the outskirts swarms of new buildings, factories and dormitories.

At 9.30 we came to Chenghsien, the junction to which we would return and from which we turned up the Yellow River. I counted 191 large new buildings beside the track, also a water-tower, a parachute-practice tower, and an electric switching station. There was one particularly large group of factory buildings, also a group of ten small steel furnaces, open and surrounded by a swarm of workers' huts. Spreading out from this town in many directions were high-tension electric lines. As we left Chenghsien I saw three large pieces of earth-moving equipment, which were the

151

only ones I saw in China. I suppose that they also has been left by the Nationalist army.

Our route now turned west and we started to climb into the hills, puffing slowly upward all day, the scenery becoming ever more rugged and strange as we entered the loess country. It was a fanciful land, built up in terraces as though it had been cut out of a series of planks. It was terraced like a mechanical draftsman's or a cubist painter's idea of mountains. All lines appeared to be straight vertical cliffs or horizontal planes save where the silhouette of some ancient and dessiccated tree broke the battlemented hills.

These strange formations have been cut from consolidated dust blown in distant millions of years from the Gobi desert. It has formed a soft rock that is able to stand up in vertical walls, so that streams cut it into steep and narrow canyons and man can terrace the hills into fields or excavate it into caves in which to live.

All day great numbers of men were at work double-tracking the railway. With picks and shovels they widened the cutting while other men with little baskets carried the fill to valleys. An idea of the number employed can be had from the picks I saw piled at one small station. They were in uniform bundles and so I was easily able to estimate their number at 2,200. Some of the baskets were carried by women but they did no shovelling. Other men trimmed stone, shaped wooden frames, or stood in a circle around a heavy stone, rhythmically pulling ropes to make it bounce up and down to tamp the earth. Small bridges and culverts were being built across streams from precast concrete shapes such as they use in Russia.

The sidings held many freight trains loaded with coal, timber, oil, and trucks, although there were no signs of any highways for the latter to run on. We passed an apparently new mine, grain elevators, a large flour mill, and many factories; and at intervals all through the country we saw electric power lines.

My attention was turned from the window to the inside of the car by Mr. Tien looking up from his reading and asking, "What is the difference in pronunciation between w-o-o-d and w-o-u-l-d?"

"None," I said.

"S-o and s-e-w?"

"None again."

"How about r-o-c-k and r-u-c-k?"

To entertain him, I showed him as many plays on English words as I could remember—the palindromes "Madam, I'm Adam" and "Able was I ere I saw Elba," and the little verse containing different words spelled by the same letters:

> *A* sutler *sat in his* ulster *grey,*
> *Watching the moonbeams'* lustre *play*
> *On a keg which in the bushes lay.*
> *"John Barleycorn, my King!*
> *To the* rustle *of leaves I tune my song.*
> *Thou* lurest *the brave, thou* rulest *the strong.*
> *To thee the* result *of battles belong,*
> *John Barleycorn, my King!"*

He read the *World News* of July 5, 1958 from cover to cover and lent it to me. The type of literature given to students in China on which to practise English is illustrated by the following quotation from it:

A colourful procession from Hyde Park to Trafalgar Square last Sunday was like the bursting forth of the sun in all its glory dispersing the dark clouds. In fact, the sun did shine on the demonstrators with unusual brillance as though to confirm their faith. So 10,000 marched with their bright banners and slogans to join the other thousands who waited them in the square where the Communist Party speakers gave them a message of hope and confidence in the power of the people to achieve the aims which were the watchwords of the demonstration—peace, work and British independence.

The Tory Government in the interest of big business is determined to force down the peoples' living standard, to create large-scale unemployment, to attack social services, housing, health and pensions and to sell us even more completely to the United States and their atom maniacs.

In an article entitled, "They Can Neither Keep Peace nor Understand," Hu Chaio-mu referred to an editorial headed "Great Leap Year" that had apparently been published in the London Times of May 31.

What a striking contrast the commentary forms to the stirring activity that marks Chinese life today! This calls to mind two lines by Liu Yu-lui, a poet of the T'ang dynasty, which read:

Past the sunken boat a thousand sails,
Beyond the diseased oak ten thousand sap-green trees.

The imperialist West is aged and decaying like a diseased oak and the sunken boat, but the socialist East is flourishing and hopeful like a thousand sails racing ahead and ten thousand trees turning green. The aged West can neither keep pace nor understand the youthful East. The West is drawing its last breath in its crumbling world.

I could not refrain from remarking to Mr. Tien: "As you have not seen the West, I understand how difficult it is for you to judge what nonsense all that is and to comprehend how vigorous and rich our system of free enterprise is; but your leaders know it and these remarks seem to me to be a little odd in view of how zealously you in the East are all copying the West."

Early in the afternoon we passed Loyang, which in a very ancient period had been for a time the capital of China. I saw only the railway station, and my chief recollection of that was of the posters. These were not printed, they were murals painted directly on the walls and much more imaginatively conceived than those in Peking. Whereas in Peking the posters tended to be rather crude and brutal, many of those in Loyang were witty and skilful political cartoons.

In Peking a typical poster might be of a vast Arab kicking, chopping or hanging two tiny khaki-clad figures draped in British and American flags—a reference to the troubles in

Jordan and Lebanon, about which it must be admitted I had had very little information.

Odd as it may seem, one of the most peaceful ways to spend a summer rent by international squabbles is to spend it alone in the heart of Iron Curtain countries. During my journey across Asia I could not read the newspapers or understand the radio. I had little contact with other Westerners who might provide information. My only sources of news were chance travellers, occasional visits to embassies, and my interpreters. I did not encourage the latter to tell me the news as I had no wish to argue with them and I knew that their opinions would be highly biased.

It thus happened that during the mid-summer weeks, when people in North America and Europe, and no doubt in Russia and China as well, were worrying about the possibilities of war in the Middle East, I rambled peacefully across the ancient land of Asia and never did learn clearly what was going on.

The posters at Loyang were delightful pieces of art drawn in gay primary colours, clearly demonstrating that the government authorities did not approve of the actions of the American and British in the Middle East. One was of a beach on a desert shore upon which two whales representing the United States and Britain had been stranded and were rotting in the desert heat. Another showed a pair of American soldiers, one of whom was distributing Cokes while his companion's head was being blown off by a bomb.

Another panel was almost filled by the palm of a great hand which was slightly cupped, with the fingers pointed straight upwards. This palm had been skilfully employed as a seat for a puppet king, and the significance of the poster lay in the canopy of crimson cloth that covered the tops of four fingers forming the back of the throne behind the puppet's head. This canopy was embroided with golden dollar signs.

Another poster showed an American soldier pulling a

skeleton out of a box with the help of his British companion; but the one with the most barbed wit was of a camel standing in the middle of a desert. Seated between the two humps was a figure, obviously Mr. Dulles, who was furiously brandishing a whip, apparently unaware that the head of the camel had been neatly sliced off and lay oozing blood in the sand at its feet.

Vicious they were if you like, but so are political cartoons everywhere. Of course they were all violently anti-Western and calculated to inflame the populace, most of whom were probably even more difficult to interest and much more poorly informed on the subject of Middle Eastern affairs than the farmers of central Canada or the United States.

The Chinese custom of building houses about central courtyards provides large areas of windowless walls, which lend themselves to this form of decoration. Little as I approve of the sentiment portrayed, I must admit to smiling at skilful paintings of Uncle Sam disguised as a paper tiger, or Mr. Eisenhower as a miserly old mandarin in a blue gown, hung with a great necklace of golden dollars that he was greedily fingering.

Besides these that dealt with political subjects, many wall posters in the various cities were advertisements. These were sometimes painted on tin billboards. They frequently had one or two words in English lettering and a considerable number in Chinese, together with a picture of the product they were advertising. Thus one frequently saw large posters covered with pictures of bottles of ink, fountain pens, radio tubes, patent medicines, farm implements, or rubber hoses and belts.

There were, of course, special notices outside movies and theatres, and at stations and along the streets were often horizontal or vertical streamers with a single row of characters.

On other walls there were pictures of the future when there would be sewing machines and tractors and when the

156

Chinese would own and drive cars. There were posters showing how to use dial telephones; one showed a group taking a lathe to pieces.

The subject of the Great Leap Forward, with its winged-horse symbol, was a favourite theme. I remember seeing this horse and rider as the nucleus of an atom, surrounded by electron orbits while they leaped into the atomic age. Elsewhere colourful pictures showed a troop of winged cavalry making the Great Leap Forward over clouds. A frequent theme depicted the Chinese advancing from oxcarts to railways to airplanes and to the promised land, often a paradise with buildings in the Ming palace style, but sometimes symbolized by a horseman in an atom.

The theme of the good harvest was a common one. It was represented by a very fat, full grain sack, which had on it a smiling face with sheaves of wheat for hair; the sack had a hole in it and was pouring grain onto the happy people who stood below with upraised hands. Several dealt with foundries out of which abundant iron was pouring, or with transmission lines providing electric power. Others emphasized co-operative effort, with large numbers of people holding hands and dancing together, or with individuals teaching others how to use the lathe or how to read.

I saw one picture of the three races of men—Negro, Oriental, and Caucasian—co-operating. In only one case did I see a poster illustrating Russians and Chinese co-operating. It interested me to observe that whereas we would portray ourselves with pink skins and the Chinese with yellow, they used pink for themselves and puce for us.

All afternoon we passed many groups of very industrious workers, well organized though they worked with primitive methods. One scarcely saw a foreman and never a soldier among them. In central China I saw little evidence of force and few troops, except in Sian and Lanchow and armed guards once at an old fort with walls and towers. I saw no signs of prisoners in the fort, and reflected that in travelling

157

an equivalent distance in North America, one might easily expect to see either a jail or an army warehouse. The fact is that I saw less sign of military activity than in central North America, and little sign of force. (I was not near the coast.)

Of course, the people are regimented, but I think it is more likely done through a system of distributing meal tickets than by soldiers. Some men looked overworked, but I expect that most of them were happy to get a full belly.

At the stations many of the male passengers got out to stretch, but like most of the others, Mr. Tien was content to stand by the door, while I tried to get a little exercise. On one excursion I was astonished to see English printing on a booklet at a news-stand and bought it for a few cents. It turned out to be a grammar or text-book, showing how to write Mandarin dialect phonetically in the English alphabet. All the newspapers also had their names written in English letters as a sub-title. It was not the conventional Wade system of transliteration, but another, so that the newspapers from various places read *Beijng* (Peking) *Ribao, Shanxi Ribao* and *Gansu Ribao*. Food and soft drinks were for sale and women tried to sell me hand-knitted sweaters in the evening as we climbed into the mountains and it grew cool.

At eight o'clock Mr. Tien said, "That is Tungkwan, the old gateway to the Chin kingdom." A few hours later we disembarked at Sian, the capital of the Chin emperors. The station at which we alighted was a modern building of grey brick whose simplicity was lightened by a few red pillars, by a tiled roof, and by a few architectural touches that gave it a suggestion of Ming-palace style.

We were met by three polite gentlemen who drove us along wide and lighted boulevards to a new hotel and ensconced me in a comfortable suite.

Sian is an ancient city, indeed one of the oldest in China, for it was the capital of the Chin dynasty, which first unified the land and from which the country takes its name. The ancient city had consisted almost entirely of low mud and plaster buildings crowded upon narrow streets. Very recently, wide modern boulevards had been driven across it and out through breaches in the old walls into the country. Often they followed former streets, but these had been widened and paved with concrete.

The Sian Hotel, at which I was to stay, was set well back from the road with gardens before it. The building had evidently been completed within the last five years and was a fine modern one of grey brick, again with traces of palace ornament. It was seven stories high, with twenty-two pairs of windows across the front of each story, one pair to each room. Behind this front part of the hotel, as I later discovered, there was a courtyard and another even newer wing that was scarcely completed. This was six stories high and had eighty windows across the front on each floor. I discovered that each block had central corridors, off which rooms opened on either side, so I estimated that the whole hotel might have seven hundred to eight hundred rooms.

I was taken to a suite in the front block, overlooking the city. The elevators, which my interpreter told me had written on them "Made in Shanghai," had also visible on the control panel the word "Otis," surrounded by a wreath; but perhaps these two labels were not incompatible. On the third floor, there appeared to be about forty bedroom doors, twenty on each side of the corridor, which validated the observation that I made from the outside.

The suite in which they installed me had two rooms and a private bathroom, but I noticed that in most rooms three Chinese were sleeping in as many beds. The sitting-room measured some fourteen by eighteen feet and was furnished with a large chesterfield and two arm-chairs, all upholstered in brown artificial leather. Beside them there were three wooden side and coffee tables of a solid Chinese design. Against the wall there was also a large office desk with drawers in it, a stiff arm-chair, and a bookcase. On the desk were an electric fan and two potted salvia plants.

The pair of double windows looked out over the city,

and by rotating handles they could be opened without disturbing the inner fly screens. These bronze window fittings worked well. There was a coil in the room for steam heat in winter, and the walls were decorated with a number of Chinese prints in picture-frames. The floor was of bare wood, painted brown with rather poor paint. The walls were gray and the ceiling white molded plaster of a simple design. There were good curtains to keep out the light.

The connecting bedroom was furnished with a bed, a bureau, a chair, and a night table with a lamp. There was a large clothes cupboard. The bedroom also had a door into the corridor so that it could be used as a separate room. Opening off it was a three-piece tiled bathroom which had hot and cold running water and plenty of towels, soap, and — pleasantly enough — a terry-cloth bathrobe.

The corridor leading from the elevator to my rooms was plain, with a patterned terrazzo floor. A red carpet to cover the corridor had been rolled to one side all the way along and was evidently used only on special occasions.

Opposite the elevator was a large meeting-room that was filled throughout the day and evening with Chinese engaged in committee meetings and ardent conversations. It was furnished in Western style, with tables, chairs and rugs. Of the two lifts in this wing of the hotel one was working, but the shaft of the other was filled from top to bottom with a scaffolding of bamboo poles and fibre ropes; it was clearly undergoing some sort of repair.

My rooms were comfortable if somewhat noisy. All over the city there was a constant humming of industry day and night, and a metal factory across the street was periodically lit by the glare of arc welding; but the city was low and flat, and whenever I looked out by day across the roofs I could see in the distance the blue Chingling Mountains.

SIXTEEN

Neolithic villages and modern communes

Friday, September 5th

After breakfast Mr. Tien and I were picked up by Professor Wang Yung-yu, Mr. Chao Wen and Mr. K'ung Hsien-chang, who had met us the night before at the station. Professor Wang Yung-yu was a plump and rather silent scientist who spoke no English. He was the professor of quaternary geology at the University of Sian and an expert on Chinese archeology.

Mr. K'ung Hsien-chang told me that he was a 76th-generation descendant in direct male line from Confucius, whose surname he naturally bears, Confucius being the Latin form of K'ung Fu-tzu. Mr. K'ung was a local employee of China Intourist. He could not tell me how many more descendants of Confucius were alive, but it appeared that there were a great many and that some were of the 80th generation. Under the old régime they were all held in so much esteem that none of them had to pay any taxes; but now they were treated like anyone else. Nevertheless, it interested me to observe that although Confucian doctrines are in the discard, his memory and his descendents are still respected.

We all got into one of the cars that were waiting with their drivers outside the hotel. Ours was a Mercedes-Benz 190 model, to which lace curtains had been added. In it we were driven east along wide, new concrete roads slashed through the city, past the railway station and across a new concrete bridge. Here the street widened to a two-lane boulevard, planted with young trees. For many blocks we passed apartment houses of plain grey brick, about four stories high, that might have been in a new suburb of any Western city.

On the outskirts of the town we came to a newly discovered neolithic village, called the Panpotsun site. We stopped at a parking space surrounded by pleasant gardens and entered the first of three or four new brick buildings in which the principal discoveries were displayed for the benefit of the public. In front of a case containing some large earthenware pots, Professor Wang remarked:

"These pots were used as coffins. We know that, because human bones are frequently found in this type of pot. You will notice that each one has had a hole knocked in the bottom of it. This was the invariable custom, and it was evidently done to allow the spirit of the deceased to escape. Clearly these people had some sort of primitive religion."

At another case, referring to the patterns on some ornamented pots, Professor Wang said: "These marks are the earliest form of Chinese characters. We believe that our writing sprang from such symbols."

We spent some time at the museum, looking at every case in turn. By the end of the tour this was becoming tedious, and looking about I noticed that the two attendants by the door were busy studying. Walking past, I saw that each had a copy of the new grammar that I had seen by chance and bought the day before, on the railway platform. These young men had pencils and paper and were practising how to write Chinese in English letters.

From the museum buildings we climbed two broad flights of steps to a barn-like canopy that covered the actual site of the excavations. Because the ground was everywhere soft loess and the weather often wet, the diggings had been protected by a large shed.

"This village," said Professor Wang, "was built on a terrace; its present height is sixty feet above the river. In neolithic times when the village was built, this terrace was a beach; but here in the hilly part of China, the land is subject to earthquakes and is known to be slowly rising." Indeed, he said, the present rate of rise has been measured and found to be about a sixth of an inch a year. The pattern of tributaries of the Wei River shows that it used to flow in the opposite direction, but has been reversed by the rise of the land to the west and consequent tilting.

We entered the shed and saw that it consisted of two main rooms, each approximately 120 feet square, connected by two or three smaller ones. Walking through these on raised wood walks, we could look down over the foundations of the excavated village.

The edge was marked by two defence ditches, each about fifteen feet deep and eighteen feet wide, and within the village. I immediately noticed the location of several former huts, marked by circular patterns of post-holes. Professor Wang pointed to other places and said:

"That hollow was a storehouse. We found pots of grain there. Beyond you can see the burial ground, for those pits contained coffins, some of which you saw in the museum."

Elsewhere there were barns and fireplaces, several of which still contained ashes. Looking across the excavation, I grasped the general plan of this ancient village that had been buried for thousands of years beside the Wei River. From this and from the exhibits in the museum, I could understand something of the simple life of the early Chinese, who had begun to lead a settled existence and who had started to write their primitive symbols on pottery, three or four thousands years before.

"Many sites like this have been found," said Professor Wang, "during the construction of new industries. This one

163

was discovered in 1952, when the foundations of a factory were begun here as part of a program of reconstruction after the liberation. It was excavated between 1954 and 1957, but we now know that beyond these buildings lies the greater part of the site, still undisturbed. I anticipate that we will start to extend the dig in 1959."

According to him, the Panpotsun site corresponds to the two latest of five stages into which neolithic cultures of the Yellow River valley have been divided. These stages, which immediately preceded the earliest Yin and Hsia dynasties, extended back from about 1,000 B.C. to perhaps 3,000 B.C. They are called the Sahchin, Sintien, Pangshan, Machao, and Yangshao stages.

Before that there are gaps, but I gathered that China is very rich in archaeological remains and that some of the more important sites of paleolithic man and their approximate ages are these:

  50,000 years — upper cave men of Choukoutien
    (which we had seen)
  50,000 years — Tsu Yang, Szechwan province
  200,000 years — Hotas culture
  250,000 years — Tingtsun village, Shansi province
  500,000 years — Peking man, Choukoutien (also seen).

"All these descriptions and dates," said Professor Wang, "are based upon work done before the liberation. So many new discoveries are now being made that it will be necessary to correct these conclusions."

What especially fascinated me about the Panpotsun site and the effort lavished upon displaying it, was the clear indication of the interest of the new régime in the historical and cultural aspects of Chinese civilization, as well as in purely utilitarian matters. I sensed the importance that the Chinese attached to these discoveries. The fact that a succession of cultures is represented in an apparently uniform series suggests that the ancestors of the Chinese people have inhabited the region of the Yellow River for thousands and perhaps hundreds of thousands of years. Thus the Chinese are the originators of their own culture; they are not copyists.

Leaving the Panpotsun site late in the morning, we drove farther out into the country, at first along the excellent boulevard, then by a dirt road to the Hua Ch'ing Ch'hi warm-spring baths, which had been built thirteen hundred years ago by Emperor T'ang Ming Huang of the T'ang dynasty for the pleasure of himself and his favourite queen, Yang Kwei-fei.

"Their love is a favourite Chinese story," said Mr. Tien. "They frequently came here to bathe from their palace in Sian."

The baths are situated a little way up the hills overlooking Sian and the Wei valley, where ancient walls and graceful trees enclose and shade a garden. Scarlet pillars of Ming pavilions and roofs of blue-green tiles are set on islands, surrounded by pools carpeted with lotus. Cut in the hillside are restful rooms of stone where we ate our lunch, gazing out over the steaming ponds and gay courtyards, mottled with summer brilliance and welcome shade. Two young artists, surrounded by innumerable pots of paint, had started to decorate one wall with a fanciful picture in honour of the new régime.

During this lunch we were entertained, along with a few other tourists and many Chinese, by an exhibition provided by a Chinese juggler and musicians, and by the impromptu dancing of a group of Australian folk dancers. By far the most active of this group was an enormous Negress, who puzzled me. Although she sang Negro spirituals with great gusto, it seemed unlikely that she was an American, and I assumed her to be a Fijian.

In general, the Australian girls were a great deal more active than their male companions, most of whom were middle-aged musicians. One of the younger men, with long, wavy hair and clad only in a shirt, shorts, and sandals, spent the afternoon sitting on a balustrade sewing the girls' ballet shoes, while the very spritely ballerinas and a few companions climbed the hill behind the Hua Ch'ing Ch'ih.

After they had departed, my Chinese hosts suggested that I should not miss the opportunity of bathing in these famous hot springs. They hoped that I would agree to do so immediately because the facilities were limited, and in this way I would avoid much congestion when the Australians returned. They said that later we could go for a walk up the hills.

Although it was a very warm day and I had just eaten a considerable lunch, I agreed, and soon found myself floating in a tiled tank seven feet square and full of hot water. At the time it seemed very delightful and relaxing, but I must say that even after a few minutes, I felt enervated and in no fit shape to climb the lofty hills behind; but my companions were most insistent that I should accompany them at least part of the way. In a very exhausted state, I presently reached their goal. This was a simple look-out like some Greek temple, consisting of a canopy resting on four big pillars.

"This monument," they said in triumph, "commemorates the spot at which Chiang Kai-shek was captured in 1936 by our victorious troops under Marshall Yang. The Marshall proposed to shoot him, but his life was saved by the intervention of none other than Mao Tse-tung himself, who said we must all unite in the struggle against our common enemy, the Japanese imperialist aggressors. In spite of his promises, Chiang broke his word and used the reprieve to further his own ends.

"As you see, the hillside is very steep and for that reason the memorial could not be built on the exact spot. The capture was effected in that crevice ten yards farther up the cliff, into which Chiang was chased by men of the army of liberation."

I could go not further. I was not disposed to argue about the accuracy of their account. I only knew that Chiang Kai-shek had indeed been captured. I sat down, mopped my face with my handkerchief and wondered if my queer feeling was because of the onset of heat prostration. Presently I recovered enough to enjoy the view out over the fertile Wei valley and to feel greatly relieved that we did not have to follow the energetic ballerinas to the top of the mountain.

On the way down the hillside I noticed that many, many small trees had been planted all over this rough ground. Of course, this was a showplace, but from the train on other occasions I noticed similar afforestation in progress in many parts of China. It was clear that their shortage of wood is regarded as serious and that great efforts are being made to grow more trees in all places too infertile for cultivation.

After a further rest and tea at the hot springs (for I was still exhausted), Mr. K'ung took us to what he called the fifth cotton mill of the north-west. Beside it was the identical sixth mill.

It was a new, reinforced-concrete building, painted white on the inside, with a saw-tooth-shaped roof giving a north light. Another Canadian, Professor Lewis Walmsley, later told me that in 1957 he had seen one of these mills being constructed under the direction of Russian engineers. I did not know this at the time, but I had looked for and seen very few Europeans in the hotel, so I assume that the Russian technicians had done their job and left.

After we had waited a long time in the mill manager's office, he appeared and answered my many questions.

"Before 1954," he said, "there was nothing in this region [ten miles from the centre of the city] except fields and waste ground. In that year we began construction, and production started in 1956.

"Three shifts are now working and producing 70,000 miles of cloth a year. In the mill are 33,000 spindles and 3,712 weaving machines, which produce five varieties of plain cotton cloth. All the machines were made in China— the weaving machines in Shanghai and the spindles in Shantung and Shansi Provinces.

167

"Last year we could not get enough cotton and so could not use all our spindles to capacity. There was also a shortage of labour, but this year we have all taken a Great Leap Forward. We have found more workers and the workers are doing better; they have speeded the work, so that each now supervises fifty-six weaving machines instead of twenty-four, or twelve hundred spindles instead of eight hundred. This month I hope that all the looms will get into operation for the first time."

"The plant works six days a week, with one day's rest. There are 5,200 regular workers, 56 per cent of whom are women. All the workers live in dormitories that were specially built along with the plant. The welfare is nearly perfect. They have dining-rooms, nurseries, kindergartens, and a park to play in. Altogether the village here houses twenty thousand people. It was built at the same time as the mill."

The pay varied according to technical level and was increased for good-quality work among workers in the same class. New apprentices received only $6 a month. For the apprentices, dormitory accommodation was free; so were health services, water, and light. Those on a higher scale paid a small amount; for example, the manager said, a room like the one in which we are sitting, which was about twelve feet by twenty feet, would cost $1.40 furnished with water and light. Unmarried people might be housed five in a room and pay twenty-five cents per month. Each received a desk and a bed.

He pointed out to me that near the gate was a large board on which criticisms of the management from the workmen could be posted. (I had noticed is as we came in, and thought that the criticisms appeared to be rather weathered and out of date.)

"That," he said, "is one feature of our way of life; everyone can criticize everyone else. In this way we advance our work with the utmost speed and with ever-increasing efficiency."

He also showed us charts in his office that compared the output of this mill with others in regard to the amount of material and electricity used and the quality of the output. It was apparent that this comparatively new mill was not as successful as many of the older ones.

We then went on tour of the plant, from the place where the raw cotton was received to the place where the final bales of finished products were shipped away. It was perfectly clear that the manager knew all phases of the work of the mill extremely well. On three occasions—once in the spinning-room and twice in the weaving-room—he stopped and very deftly tied threads or corrected a fault in the machinery. On one of these occasions I pointed out to him that some confusion was arising and he immediately knew how to fix it. He struck me as a very accurate and able man.

Everywhere there was new, modern equipment. I am no judge of cotton mills, but it seemed to me that it was efficiently planned and well built and that the workers were very busily employed, but I did not know whether in all cases the operations were as safe as they might have been. The noise in some parts, especially in the room that contained all the 3,700 looms, was absolutely deafening—it was almost impossible to make oneself heard.

Propaganda was very widespread throughout the mill. On many of the looms vertical wires carried little banners on which slogans were written. All over the vast room triangular green, pink or purple flags bobbed up and down with the movements of the looms, adding a touch of gaiety to what might otherwise have been a dull and mechanical scene.

All around the walls there were many cartoons. Some illustrated the great progress to be expected in transportation, showing the advance from men acting as beasts of burden, to carts drawn by oxen, to motor cars, to jet airplanes, and finally to the heaven of the future. There were many exhortations to work harder during the Great Leap Forward of 1958.

169

Outside the mill we went immediately to one of two identical nurseries that lay on either side of the main entrance, as close as possible to the work. As we walked through the corridors, we passed rows of nursing mothers, sitting on stools, feeding their offspring as we passed. In the rooms off these corridors were more howling or sleeping infants.

"These nurseries," said the manager, "are reserved for children under one year of age and are purposely placed close to the work, so that the mothers can come out for short intervals during the shift and feed their babies."

Across the road and a little further away were other day nurseries for older children. Here great groups of twenty, thirty and forty children were being tended by young girls. They all seemed quite happy and healthy. They were not as numerous as I would have expected, but apparently children were placed there only if there were no older sisters or grandparents to look after them at home.

Beyond we came to the workers' living quarters, which were of two types. There were large dormitories for single men and women, and there were row-houses of one room each for families. Each room had a door and a window in front and behind. They varied a little in size and by the doors were often a few sunflowers or young trees. Around these houses the families were going about their business— playing with their children, doing the family washing, or merely sitting and resting in the evening sun.

We did not enter, but we saw the higher apartment buildings in which the dormitories for single men and for single women were housed. They were cut off from the family dwellings by wire fences, and it was explained to me that this was not from any desire to restrain the inhabitants but merely to keep the multitudinous children from being a nuisance, which seemed to be sensible, for there were children running about everywhere else.

In the midst of these dwellings were the community kitchens and dining-halls, through which we walked. At the time there were comparatively few people eating, but it was clear that abundant quantities of simple food could be had from the large kitchens. One feature was a series of a dozen taps on the outside wall, supplying boiling water. To these a constant succession of women and children came with thermos bottles from the houses. It was evident that when they wanted to make tea or wash the dishes, they came to the kitchen and got a free supply of boiling water.

On the ground around the dormitories very large numbers of earnest young men and young women were sitting in groups, I presume discussing Communist philosophy and the program of the Great Leap Forward. As we were leaving the company town, we went down a street with two or three motion-picture theatres and a number of small shops that had the appearance of being private, although I do not know, of course, how they were organized. Also drawn up here were buses that operated regularly into the city; and a few of the wealthier or more energetic of the mill workers were riding bicycles. It was clear that they had freedom to come and go between the village and Sian.

As I thanked the manager, he asked if I had any more questions.

"Yes," I said, "how many holidays do the people get a year?"

"Fifty-six days."

"Does that include the Sundays?"

"Of course," he said, quite seriously; "but if families are separated, they can get one week a year of separation leave, or even more if they have far to go."

We might regard such a life, of continuous work with a 48-hour week, a one-room home and simple food from a community kitchen, with abhorence as a mild but dreary form of serfdom.

Not so the Chinese. As we drove away, Mr. Tien, who had a very hard time during the civil war and who had faced starvation and little but work all his life, remarked in all seriousness, "It must be nice to work in one of the cotton mills."

I looked at him questioningly.

"So good to have such a steady job."

I had just been through a commune. No doubt it was one of the better ones, but not unique. The Chinese have to produce cotton cloth for 650,000,000 people. Why would they not seek to do so efficiently? This commune, so far from being regarded with horror by the average Chinese, was from the point of view of many of them a desirable and luxurious way of life. Instead of a mud hut, they had a brick room; instead of farming from dawn to dusk, they had an eight-hour day; instead of the threat of starvation, there was a community kitchen.

Older people, intellectuals, the religious, and those who were formerly well-off, might resent this regimentation, this disruption, the lack of privacy, the continual pother over politics; but the coolie who had had nothing before liked a full belly, and the young mother who would have had to work anyway appreciated a day nursery and a dining-hall.

Perhaps what they all liked best of all, after forty years of civil war and invasion, was security. It seemed odd that the search for security, so often decried as the disease of modern North American youth, should be so rapidly becoming entrenched in this Communist state.

We drove back to the hotel and had a normal supper, consisting of four main dishes: chicken with peas, green bamboo shoots, cucumber and fungus soup, and meat cut in little, thimble-like pieces covered with spines like a hedgehog. I inquired about this meat, and Mr. Tien said it was a special delicacy that only a few people knew how to prepare. When it was all finished, the waiter brought in a great wooden box, a one-foot cube like a beehive, and steaming hot. It was made up of a number of trays with perforated bottoms, on each of which were rows of steamed pancakes stuffed with meat. There were some thirty of these delicious pancakes, of two varieties. I am ashamed to say that Mr. Tien and I managed to eat twenty-two of them before we tottered off to bed.

SEVENTEEN

Of Confucian history and confusion of history

Saturday, September 6th

The European dining-room on the top floor in which Mr. Tien and I had breakfast was divided by movable silk screens into a number of cubicles, so it was difficult to tell who else was present, but as far as I could make out, there were only some Australian folk dancers, two French women and an Englishman, who remarked that he was going by the new perilous railway to Chengtu which lies to the south over a 12,000-foot pass. The Chinese ate in a much larger dining-room of which we only had glimpses.

After the strenuous day yesterday I found it hard (especially with chopsticks) to eat pickled vegetables, dumplings, black eggs and rice stewed to a soup. The black eggs that one often hears about have been so treated that the white is a dark brown jelly and the yolk a greenish paste. They are highly prized for their taste, which to me suggested the bastard offspring of a hen and a pickled mushroom.

While we were eating, Mr. Tien aroused me by remarking, "It is very important that a people should be active. Which do you consider the liveliest race, the Americans or the Australians?"

It was not a subject that I had ever considered—or thought important that morning—so I put him off by replying, "I

174

don't really know, but I think that the Chinese are among the most polite."

"The Chinese," he said, "'follow Confucius too closely. Confucius said, 'A man should be quiet and not say much.' We Chinese are now trying to live this down. We are trying to be more lively."

On several other occasions he asked me about this liveliness of people; it was clearly a matter that occupied his mind. He next astonished me by saying, "It is too bad that it is so hard to get enough workers for the mills. Although the young people are encouraged to move around the country, the labour shortage is still very serious."

"You can't possibly believe that!" I told him. "One of China's main problems has always been famines, and the population is now increasing so fast that it must be difficult to feed them all."

"You don't realize," he responded, "how inefficient China used to be, nor do you appreciate the great improvements introduced by our Communist Government under Chairman Mao. Consider the 1958 harvest; it will be far greater than any before. Our production of grain has recently doubled, and when the new irrigation and flood-control schemes are completed there will be plenty for everyone. At the present time our vast program of construction could employ many more people than are available. This is in spite of the great increase in the labour force made possible by the emancipation of women. Formerly, following Confucian doctrine, women were confined to their homes. Their sole duty was to serve their fathers, their husbands or their sons. They are now free to work in the mills and fields."

After breakfast we drove to the Northwest University on the opposite side of Sian and were met by the vice-president, Professor Cheng Po-sheng (a distinguished and charming scholar); Professor Wang Yung-yu, who had accompanied us the previous morning; two grey-headed, stalwart geologists,

175

Professors Chang Pei-shen and Mo Shi-chen; and the head of the physics department, the intelligent, nervous and be-spectacled Kiang Jen-shu.

After polite introductions, we were ushered upstairs in the main building of the university to a plain room that might have been a meeting-hall or a faculty dining-room, for it held a U-shaped table covered with a white cloth at which sixty-five people could sit. In the corners were pots of flowering plants, and on the walls an electric clock and pictures of Mao and of famous local buildings.

When we were seated with our tea and cigarettes, the vice-president explained to me the history and work of Northwest University. (He was kind enough to allow me to take these notes while he was speaking.)

"At the beginning of the Sino-Japanese war, three universities that had been located at Peking were moved to Chengtu and became united into Northwest University. At the end of the war the university was divided into medical, teaching, engineering colleges and so on, some of which came to Sian. Some of the university professors left Shansi for Peking. Thus, before the liberation the staff was weak and the buildings poor. Since the liberation, new buildings have been started; facilities and instruments have been provided. The teaching staff has been increased, so that now there are fifty professors and another two hundred and fifty junior instructors, but because the university is so newly established, the proportion of new and inexperienced teachers is high. The number of students has increased from seven hundred to three thousand.

"After a period of reorganization, there are now ten departments: mathematics, physics, chemistry, biology, geology, geography, law, economics, history, and Chinese literature.

"In general, class-room education is combined with practice. For example, at the present time many geologists are making maps in the country; students of geography and

176

biology are in the mountains exploring; students of economics are working in the factories, and only the freshmen are to be found in Sian when the term is about to begin. This year there will be nine hundred new students, but only three hundred are coming from Shansi, the rest from all over China, mainly from Honan and the Yangtse valley. Although the number of high schools in Sian has been increased from four to twelve, and six more are being built as part of the Great Leap Forward, there are not enough high-school graduates to fill the four universities now located at Sian — this one, Northwest Industrial University, the College of Commerce and the College of Architecture. We have therefore brought students from elsewhere.

"Students get free tuition, dormitories and medical treatment, and at least 70 per cent of the students get subsidies or scholarships to pay for food and clothes."

"As at the University of Peking?" I interjected.

"Yes," he said, "but of the senior middle-school students, only a quarter qualify for college; the rest have too poor marks. Any that qualify can go to college.

"The students have some choice of courses. The decision is based partly upon preferences and partly upon the results of entrance examinations taken upon arriving at the university. Thus if the physics department plans to take one hundred and twenty students and three hundred students wish to enter, the one hundred and twenty students with the highest grades are accepted and the other hundred and eighty have to be content with their second or third choices. The university course is four years and, in the College of Commerce, five. As yet we have no graduate work.

"After graduation, the method of placing students is democracy mingled with centralism. There is a general shortage of university graduates, and of course every ministry and academy wants them and sends lists of requirements. Each graduate writes a report about his wishes and then in turn takes part in a public debate. After discussion, it is decided

177

to what place each student is to go. The university does its best to meet the desires of the students. For example, if a man has to look after his family, he may be certain he will be sent to a position near his home; but if he has no family he may be called upon to go farther away. Ultimately, the school authorities will persuade students to go to the best places."

A bell at 9:30 brought some of the students to the courtyard to exercise. They did not appear to be very enthusiastic about it, but after five minutes quite a number were doing calisthenics. This reminded the vice-director to point out that care was taken of the students' health and that less work was given to weaker brethren.

"Every year the university has a new building program which is regulated according to the growth of the student enrolment; this year, for example, buildings to accommodate five hundred more students are being added."

The physics professor then asked me to explain what we were doing in physics at the University of Toronto. After I had replied, he told me that the physics department wished to exchange literature, particularly in the fields of low-temperature physics and the viscosity of metals. He said, "Although we have no graduate students, we professors do research."

We then toured the campus, starting at a meteorological and seismological observatory that was contained in three rooms. There was a receiver for radio time signals, a clock made in China and a room devoted to interpreting earthquake records. I went down some stairs into an underground vault and saw two sets of horizontal seismographs—mechanical recorders of the 1951 Type 1 and Type 3 models—for recording the large earthquakes that occur in the Yellow River valley. In a third, unlighted, room were three Chinese Kirnos instruments. On the roof were ordinary meteorological instruments.

Outside, a new library building, five stories high and perhaps two hundred feet long, was nearing completion. It was completely surrounded by the traditional type of bamboo scaffolding, tied with ropes, but I noticed that an electric hoist instead of man-power was being used to pull up materials.

Another new U-shaped building, which we did not visit, was said to house biology, geography and literary studies.

Physics and geology were housed in a typical well-built, three-story grey-brick building. Most of the few students in it were busy putting up wall posters, written in Chinese characters on old newspapers and on green and purple paper.

On the ground floor were physics laboratories, many of them equipped for advanced, if old-fashioned, experiments in optics. The vice-president, himself a geologist, remarked that optics was the president's specialty. The instruments had been made in many countries. There were optical instruments from Zeiss in East Germany ,electronic equipment from Hungary, chemical balances from the United States, spectrographs from Japan; but most of the common meters, resistances, batteries, cathode-ray oscilloscopes, and common electrical parts were made in China; so was one big arc generator. I asked whether they had much Russian equipment, but they replied that the Russians were busy. On the walls I noticed tables of the technical elements and other charts printed in Chinese. The laboratory had many pictures of Russian scientists and of Mao Tse-tung on the walls. The only books that were actually in use by the few students there were in Chinese or English. The library had about 150 periodicals divided into Chinese, Russian and Western language sections; the last were mostly in English, but there were also some Italian, French and German periodicals, and I noticed the *Handbuch der Physik*. While showing them an article in it, I discovered that it was a multilithed copy; so were the copies of *Nature*. When I remarked on this, they became confused.

There were a few men and women working in every laboratory, and there was a meeting in the physics library. When we entered, all the participants politely stood up. Somehow it gave me the impression of being a political meeting rather than a technical one.

On the next floor I visited two large rooms devoted to a geology museum, and containing a more academic selection of specimens than I had seen in the Institute of Geological Prospecting. On the walls there were charts in Chinese illustrating evolution, Peking man and progress under the new régime.

They had a good geological library in which I found long series of such standard journals as *Economic Geology, Bulletin of the Association of American Petroleum Geologists,* and the *Journal of Geophysical Research.* I noticed that *Economic Geology* was in its original form up to 1956 but that subsequent issues were multilithed. When I again asked about the change in style, they remarked that it was hard to get Western periodicals. They had masses of Russian journals, including multilithed back numbers of *Transactions of the Academy of Sciences of the U.S.S.R.* All the professors understood a little English, but clearly they were for the most part unpractised in speaking it; they mentioned that although there were no departments of foreign languages, all graduate students in physics were expected to read English, Russian and German literature as well as Chinese.

After a brief cup of tea, we said our thanks and drove back to the hotel. On the way we saw a number of soldiers in a square; most of them were watching a few of their number in gym suits performing on parallel and horizontal bars.

After lunch we drove to the ancient and historic temple of Confucius I had heard so much about. On the way Mr. Tien told me of the poor opinion in which Confucius is now held, saying, "Formerly, the ruling classes used Confucian philosophy as a tool to maintain the status quo. They used his doctrine, 'Son should be like father, king like king and subject like subject,' to protect their own selfish interests and to suppress the masses."

Nevertheless, the ancient temple was well maintained and we were cordially received by Ho Chih-cheng, the charming old secretary and custodian. The museum, as it now is called, is arranged in two parts: the Confucian temple, founded a thousand years ago in the Sung dynasty, and an annex built in 1936 to protect Confucian relics. Both parts consist of several buildings set in a garden. The place is now intended to be a popular museum and I felt that this objective had been achieved, for the displays were simple, not too numerous, and understandable by the illiterate. They included systematic collections of Chinese pottery and of bronzes from the early Han dynasty to the present. The things that particularly struck me were some beautiful porcelain T'ang horses, a set of twelve porcelain animals, each a caricature of a courtier, and some huge bronze vessels, dating from 1,400 B.C. There were marble graves decorated with the phoenix, suggesting resurrection, and with the simple symbols from which modern Chinese ideographs have evolved.

Among the modern exhibits was a bed of the most ornate description used by the Dowager Empress only fifty years ago. There were maps, showing the travels of the pilgrim Fa Hsien, who in 399 A.D. went overland to study Buddhism and returned in 416 A.D. by sea via Ceylon and Singapore; and of the monk Hsuan Tsang who went to India by the Khyber Pass in 600 A.D. Throughout the exhibits were many examples of Indian style, particularly in the repetition of one figure many times over, evidence of the exchanges which have taken place for at least two thousand years between these two countries.

A diorama of life in Han times that portrayed two classes of people — the wealthy and their servants — was followed by a picture of the Peasant Revolt against these conditions, which took place in 100 A.D. in Sian. Statues were there of the early historians of China; the first was Szu Ma-ch'ien, of whom Mr. Tien remarked blandly, "His writings are better as literature than as history."

As in Peking, they had a model of the first seismograph. In a display devoted to Huang, who invented the compass about 233 A.D., there was a model of a compass cart, the wheels of which were connected by wooden gears in such a manner that they beat a drum once every mile. A map showed the growth of paper making from the time of its discovery in 12 B.C. in China until it reached Europe in the 16th century. It slowly followed the Silk Route, reaching Egypt in approximately 900 A.D.

The temple was founded by the T'ang emperors to honour their ancestors, but many of the rooms and separate buildings date from the Ming period. Behind are the modern houses built for the great antiquities — the world-famous stone books. Two halls contain the gems of the collection. These are one hundred and fourteen huge stone slabs, each eight feet high, on which are engraved the thirteen canons of Chinese classics, certain of which were supposedly collected and edited by Confucius himself. These thirteen canons concern history, natural sciences, manners and behaviour, poetry, the sayings of Confucius and a dictionary. The original stone carvings were made for Han Emperors, but by the T'ang period there were eight sets. All of them have since been destroyed or lost except this one, which was carved during the T'ang period in very beautiful calligraphy with no less than 600,000 characters inscribed on both sides of the black slate tablets.

Scholars used to learn the complete text by heart. This is said to be a perfect text without a single error; this perfection is easier to understand when one realizes that any sculptor who made a mistake was executed. In other buildings there are many other fine examples of calligraphy all carved on black slate. There is a four-sided column adorned by a poem written by the Emperor Ming-huang of the T'ang dynasty. Huge stone turtles bear on their backs vertical slate columns, eight and even ten feet high, engraved with poetry. One slab was pointed out to me because the story on it concerned a branch of Christianity, but it must have been a relatively unorthodox sect, for the secretary remarked that its adherents did not believe in the Trinity. I have learned since that it is the famous Nestorian Tablet.

Mr. Tien was enraptured by the beauties of the carving and the care with which the letters were fashioned; some were primitive forms and others were "grass" writing which the Chinese used for speed. Apparently all can be read today.

With special emphasis the secretary showed me four T'ang horses, dating from 700 A.D. They were about half life-size and carved in deep relief.

"Two others," he said angrily, "were taken by the Americans to the United States in 1914. In their efforts to obtain another, they broke it, and these fragments are all that remain. Here are photographs of the two that were stolen and you can see how beautiful they are." I wondered what the truth of the matter was.

Near the conclusion of our tour, we stopped for tea in a charming room, light and airy, furnished with soft chairs, side tables and a small rug. A shaft of sunlight swept back like a curtain from the open door which revealed outside the peaceful gardens and groves of tablets. Two aged Chinese gentlemen, so bent that their thin beards touched the sticks upon which they leaned, pottered past, and children quietly playing by the door looked in at us from time to time with curious faces. A queer buzzing sound from one of them puzzled me until I found that the youngster held a live cicada insect in his hot little hands, which in spite of imprisonment buzzed intermittently in the clear summer weather. It was very peaceful as the charming old scholar, our host, poured tea and spoke about his treasures. On the walls of the tea-room were two rubbings of old slate carvings, showing Buddhas surrounded by delicate foliage. There were photographs of the two horses that had been taken to the United States. On the tables instead of flowers were bowls containing corals, curious eroded pieces of limestone, and porcelain

183

decorations. A guest book was produced, and I was asked to write.

I saw that during the past several years, the inscriptions were few—not more than one every few weeks—and, with one exception, all were Chinese, Indonesian, German, Russian, or other languages that I could not comprehend.

I wrote cautiously, "As the first Canadian scientist to be officially received in China for eight years, it is a pleasure to acknowledge the generous manner with which I have been welcomed. It is indeed a privilege to have had glimpses of the great history of this ancient land as well as to see so much evidence of the progress which the Chinese people are making in science. J.T. Wilson, University of Toronto, President, International Union of Geodesy and Geophysics, 6th September, 1958."

I felt honoured when I saw the inscription above mine, beautifully written by a great scientist and scholar although one with whose political views I doubt if I would agree:

"7th July, 1958—Thirteen years ago I first visited the celebrated museum of the city of Sian; it was a gloomy day and the paper flapped dolefully on the windows of the dilapidated buildings, housing the famous stones of the city's Pei Lin, the 'Forest of Monuments'. Next door the old Wên Miao was shuttered up among its courts, overgrown with weeds and the filth of an army's stables. As the *Tao Tê Ching* says, 'Where an army has passed the ground is thick with weeds.' Today, in brilliant sunshine, the new paint on the Confucian temple and the Pei Lin building is resplendent, and they contain a splendid historical teaching museum as well as the wonderful old stone steles and tablets. It was a pleasure not only to study them but to see the ordinary people of Sian appreciating them too. Is this not an important aspect in the liberation of this great country and culture? Does it not guarantee that the Chinese people will for centuries to come draw inspiration of the great treasures of their past? Joseph Needham, F.R.S., University of Cambridge, President Britain-China Friendship Association."

At supper the *pièce de résistance* was a chicken (head and all) flattened in a press until it was as thin as a pancake, roasted, faintly spiced and served with green walnuts. After we had dined, we set out to visit the Chinese circus.

Professor Wang, Mr. Tien and I drove through the dark city. As we passed a modern market-place Mr. Tien said, "Formerly that was a place for prostitutes. There are now none in China. They have been abolished. So have drug addicts and beggars"; and indeed I never saw any, but I do not know what became of these unfortunate people.

The theatre, which held about a thousand people, was full of ordinary Chinese. Crowds always looked the same—there are no degrees of elegance in cotton work clothes. I noticed many children, only one soldier, and no other Occidentals. The only car outside the theatre was our own. Inside were many paper posters written both in Chinese and in English letters, for example, *"Maozhuxi Wansui!"* which literally means "Mao, Chairman—Long Life!" There was also *"Zhongguo Gonchandang Wansui!"* meaning "Chinese Communist Party—Long Life!"

The Chinese circus was a first-class vaudeville show with twenty acts, mostly juggling, and no animals at all except for a lion dance which proved as popular a finale as had a similar act in Peking. Every kind of juggling, balancing, sleight of hand and trick bicycle riding was included, and I was especially impressed by the conjurer who stood alone in the middle of the stage and produced no less than seventeen glass bowls full of water, some with goldfish, from his mandarin cloak, and by two plump and cheerful middle-aged men, each of whom balanced a heavy earthenware crock sideways on his bald and polished head. These they rolled back and forth on their heads, spun and tossed from head to head as though is was the easiest thing in the world to catch a big pot on the top of one's head and have it rest there on its side without rolling off.

On the way out I noticed a blackboard at the back of the

185

room on which somebody had inscribed the English alphabet and Chinese characters. Apparently classes were conducted there by day.

Back in my room over a cup of tea and an apple, Mr. Tien remarked, "The United States fleet is supporting Chiang Kai-shek. Formosa is part of China. We do not like U.S. interference in internal Chinese affairs."

I had no wish to get into a fruitless argement but I felt impelled to uphold the Western point of view and to reply to so direct an attack. I said, "The Americans have not the least intention of attacking China. They only want to stop Chinese expansion."

"It is American imperialism. The Morgans and Wall Street want markets. The Chinese are not imperialists. We only want peace, friendship and brotherhood. It would be a splendid thing if people could live together without exploitation."

"I agree," I responded, "and it should be easier for people to understand each other's point of view now that jet airplanes are being introduced to make exchange easier."

"The circus we have just seen," he said, "has been to twenty different countries."

"Did you enjoy it?" said Mr. K'ung, who had just joined us. "Did you think it was good?"

"Yes, excellent. The best juggling I ever saw. I am glad it has been abroad because exchange of artists, athletes and scientists are possible without political nuances and they can do much to reduce tension and show that people everywhere are human beings with similar talents, fears and desires."

"I am glad to hear you say that," said Mr. Tien. "We have to raise the standard of living of our people. We merely want to be left alone and not to be attacked by American and British imperialists."

I replied, "It is strange that you should talk to me, a Canadian, about American and British imperialism. Do we not live next to United States? Were we not once a colony of

Great Britain? We would be fearful of American domination if it were a threat, but we have not suffered much from it; and as for British imperialism, are we not now a free country? No countries get on better together or do more trade. If you Chinese complain about allegedly imperialist countries while we Canadians who are so much closer to them do not, do you not think that you might yourselves be partly to blame for your difficulties?"

Ignoring this he spluttered, "Why do the British and Americans send soldiers to Lebanon and Jordan?"

This was tough for me because I didn't really know any details of what had been going on, so I replied, "I am not clear why this should concern China. If countries care to have an indigenous Communist government that is their own affair. What we in the West are opposed to is the expansion of international Communism by subversive methods. The Hungarians uprising has shown how unhappy subject people are. Incidentally," I continued, "I have been rather surprised not to see more Russian influence in China."

"The strength of Russian influence in China is a false idea planted by American imperialists," he replied hotly. "And in any case foreigners should not interfere in these matters because bad governments are thrown out by the people."

"What you say is wishful thinking, Mr. Tien. As we saw illustrated in the museum this afternoon, China has had oppressive governments at least since the Han dynasty. It took a long time to throw the emperors out—at least 2,000 years."

He ignored this and started another line. "China's troubles began with the Opium War in 1840. The greedy British imperialists tried to turn us into dope addicts in order to get a market for their Indian opium."

Here he had me again, for I had forgotten what little I ever knew about the Opium War. No doubt it was a discreditable affair and as such not dwelt upon in our history books. So I said, "I admit that we Westerners in our dealings with China did not always behave well. Merchants advanced their own interests and our governments backed them up, but you should nevertheless be grateful to them as well as to those who acted unselfishly. Had not the West forced its way into China, you would still be serfs under a decadent Chinese imperialism. Very wisely you are copying us in an attempt to raise your standard of living. Every dynamic innovation introduced by your new régime is borrowed from the West. You have discarded the passivity of Confucianism, which now you so deplore, for Marxism, a political philosophy which was concocted by a German in the British Museum. You have copied the Industrial Revolution from Britain and adopted mass produciton from the United States. For Chinese ideographs you are substituting a Latin alphabet. Your clothes, your haircuts, your emancipation of women, your sports, all are copied. Much as I deplore some of its manifestations, I congratulate you upon the spirit in which your country is emulating the West."

"What you say may sound superficially correct," said Mr. Tien, "but of course it is wholly wrong. However, the hour is late. We should go to bed and get some rest. We can return to this discussion on the train tomorrow." That was the last I ever heard of it.

As he turned to go I inquired what the racket was in the streets. It had been steadily rising in volume until it now drowned out the usual cacophony of hammering and arc-welding from the factory opposite.

"I expect it is the people coming out of the motion picture theatres," explained Mr. Tien. "There are now over thirty of them in Sian."

I thought that the picture must have won a Chinese Oscar, or else Mr. Tien's ambition for a livelier Chinese nation had been achieved, but I let the matter pass and went to bed.

Half an hour later I was up again hanging out the window to watch a procession of several hundred Chinese, waving flags, beating drums and shouting slogans, as they passed down the main street below me. Noise from other quarters suggested that it was only one of several parades. The din continued until two in the morning.

EIGHTEEN

An outsider at 100,000 to 1

Sunday, September 7th

In front of the hotel, Mr. Tien and Mr. K'ung walked past the Mercedes Benz, a 1958 Plymouth, an old Dodge, some Pobedas, and an Austin to usher me into a Zim, the Russian equivalent of a Buick. In it we drove south for two miles along a boulevard that was being built along an older road, straight to the Great Pagoda. It is an imposing tower, seven stories and 208 feet tall.

"The Great Pagoda was built in 652 A.D.," said Mr. K'ung, "by the second T'ang Emperor, Li Shih-ming. He built it in honour of the famous monk, Hsuan Tsang, to designs which the latter brought back from India. It was the first large pagoda to be built in China.

"Hsuan Tsang had long wanted to travel to India to study Buddhism, but he was not allowed to leave China. However, about 700 A.D., a famine provided him with an opportunity to escape into Sinkiang. There he nearly died of thirst before he was rescued by a local king, who gave him money and attendants, so that eventually he was able, by going west of the Pamirs, to enter India by the Khyber Pass. There his reputation was already well known and he was graciously received, being placed on an elephant and officially welcomed in a parade attended by a half million people, so it is said. He quickly learned the five principal languages of that part of India and had discussion with the great scholars, who recognized him as their superior in learning. After seventeen years he returned to China, bringing back with him six hundred different sets of Buddhist classics. On his return, the Emperor welcomed him and built the Great Pagoda as a place where he could spend the rest of his life translating the great Buddhist classics into Chinese.

"A festival is held in his honour from the 14th to the 16th of January each year, and 200,000 people attended the last great gathering in 1946. The queue of people waiting to climb the Pagoda was over three miles long."

Nobody mentioned how many pilgrims came in 1958.

We then climbed up the Pagoda ourselves, but since there was hardly anyone there we did so with no more delay than was necessary to catch our breath. It was a stalwart tower with a fair-sized room on each floor. From the top of the Pagoda, we had a good view of the surrounding countryside. I was not allowed to take photographs, but I could see all the buildings we had visited during the past two days. According to the plan, they told me, Sian will be completed in another twelve years, but the university will be ready before that. Many dormitories are to be built on new concrete roads like the boulevard to the Great Pagoda.

At the foot of the Pagoda was a small garden, and workmen were dismantling buildings and tearing down mud walls to extend it into a large ornamental garden to the south and east of the tower. There was already a nursery of trees being raised for the purpose.

In the garden was an ancient temple containing Buddhist statues. I could see one old woman at worship and a few pitifully small offerings of fruit and flowers. In the tea garden to which we repaired were several elderly gentlemen in ordinary blue shirts and trousers. I was briefly introduced to one of them, with the explanation, "He is one of the remaining six monks." There was also a huge bell hung in the grounds, which they said weighed 20,000 catties. (A catty is about a pound.)

Returning to the city by a less direct route we passed close to the Little Pagoda, damaged by an earthquake in 1555 that had fortunately not affected its taller and much broader mate.

On the outskirts of the built-up area we began to encounter parades in increasing numbers until in the centre of the city the streets were filled with people. Some were going in one direction, some in the other, so that it was only with patience and by skilful negotiation that our driver got ahead at more than a walking pace.

The crowds around the car on every side, although all civilians, were drawn up in orderly fashion like soldiers in column of route—there can be few people more docile or easily dragooned than the Chinese. They were mostly young, and many were children parading by age groups as though the school classes had all been marshalled. Practically everyone carried a little paper flag on a bamboo stick, but each group had a different colour. At intervals, at a leader's word of command, they raised them in unison. It was gay to see all the tiny paper triangles fluttering in the sun, rising and falling over the crowds of black-haired heads that surrounded our car like waves on the sea. Many of the larger processions were spearheaded by a few big banners held aloft and were often accompanied by a large drum or gong on a cart. Two or three sweating cadres stood on each cart to beat the drums and bellow slogans in solo part, which the masses would repeat in chorus. The columns looked like so many caterpillars, furry with pennants, with banners for antennae, and with all the noise, brains, and activity concentrated in their heads. The bodies crept along behind, marching when told, shouting slogans when told, but otherwise content to bask in the sunshine of a gala day.

"You see the spontaneous expression by the Chinese masses of their hatred of the American imperialist aggression."

192

said Mr. Tien. "The Chinese people have risen as one man to repel the United States invasion of our country. We will defend our homes and repel the American invaders and liberate the oppressed people of Taiwan from the corrupt dictatorship of the Chiang Kai-shek clique."

"Who did you say was invading whom?" I said. "I am not sure I heard you correctly. Forgive my naiveté, but in my ignorance of political matters I am a little confused by what you say. I can't follow what makes you think a parade of 100,000 people, each with an identical flag, can possibly be spontaneous."

"Yesterday evening," he continued, "Premier Chou En-lai told us over the radio of the bombardment of Quemoy by the wicked imperialists who hold some of our country in subjugation. As soon as our people heard that, they rushed to the streets to demonstrate. Today the Chinese people have risen as one man in the defence of their brothers who are oppressed in Taiwan. What right have the Americans to support the Chiang Kai-shek dictatorship? There is only one China!"

By this time we reached an old Ming gateway on the new concrete boulevard that led past the hotel. The monument was undergoing restoration and the road broadened and swung around it in a traffic circle.

"Perhaps you would like to take a photograph of this famous relic," said Mr. Tien, and he stopped the driver and we got out. This stretch of road had been cleared. People lined the sidewalk peering in the direction of the sound of an approaching band. Quickly I got my picture and stood waiting on the curb until around the corner marched a military parade, led by the Chinese version of an army band and consisting of a battalion of troops in khaki. None of them was very well drilled and only the first company, with sten guns, had any arms at all. As they marched they shouted slogans, and I was interested to observe that the leaders of the chanting were neither the officers nor the sergeants, but an occasional man in the ranks who was marked out from his fellows by his purple, perspiring face and sometimes even froth around his mouth from the vigour of his exertions.

This was the great "Hate America" campaign. I subsequently learned that similar parades were marshalled all over China on this Sunday at the height of the Quemoy shelling. A whole series of these hate campaigns have been launched at regular intervals to unify the Chinese and spur them on to greater efforts. At the time I was there I saw and heard nothing of the earlier germ-warfare campaign or of the campaign against dogs, but there were still a few posters left from the war against the four evils—sparrows, flies, mosquitoes and bugs. There I stood in the midst of the latest one. It had obviously been planned for weeks and arranged to take place on a Sunday so as not to interfere with work.

I recalled that at dude ranches in Montana, where I had bummed meals as a graduate student doing geology in the mountains, they had a favourite saying, "Make them tired and you will keep them happy." So the dudes were routed out for early morning rides, kept on the go all day, and left with no time or energy for complaining. This was clearly the formula the Communists were using to deal with the Chinese people. They are made to work very hard, and all their spare time is occupied with occasional parades and frequent and futile discussions that invariably end up in agreement with the party line. Coupled with the hate campaigns was a vigorous policy of propaganda in favour of the new régime and its achievements, and a succession of campaigns of a positive nature—clean up China, abolish illiteracy and increase production in the Great Leap Forward. In this way a lot of reconstruction was being accomplished, everyone was much too tired and busy to organize an opposition, and the peoples' minds were well occupied so that they had little time to feel the loss of their old religions and philosophies or the disruption of their lives.

There is no doubt that all this is diabolically ingenious. It seemed to me to be about as effective a way of pulling a country up by its boot straps as could be devised. Concerning its impact, I can only say that I distinguished three classes of people: the militant communist cadres who formed the head and brains of every parading caterpillar, the youthful followers in the parade who went along with the gaiety and excitement, and the masses on the streets who lined the road or inscrutably went about their business on the pavement behind. Certainly no one in the parade looked resentful or obviously opposed to it. For the most part, the enthusiasts were the younger people who form the majority of the population; they have more to look forward to if standards can be raised, and less to regret if the familiar past is lost than the older people on the sidewalk. The bystanders paid less attention to the parade than those in a Western country would have done. It was part of the new China. It was fate and they accepted it.

What surprised me most was the lack of any sign of hostility or resentment to me personally. The other Europeans in the hotel had left, so that as far as I knew I might have been the only Westerner in Sian. Here I was standing on the curb at a principal intersection, head and shoulders above the Chinese crowd, our position marked by the one car parked for blocks along the street, and no one appeared to give a tinker's damn. Nobody looked at me questioningly; no one jostled me. Perhaps no one realized that I was a Westerner and not a Russian, but I suspect that the ordinary Chinese is by nature polite, far from bloodthirsty, and rather apathetic.

After lunch a little party of professors came and politely escorted Mr. Tien and me to the station where we boarded the express for Lanchow.

I dozed through the lazy afternoon after the rush in Sian, awakening at stations only to write a few notes or look at the scenery before falling back to sleep.

On the platforms were copies of Sian's daily newspaper,

the *Xi'an Ribao*, with a map of Formosa and the straits, and a cartoon of General Eisenhower and Mr. Dulles as arsonists lighting a fire. According to Mr. Tien, the headline read, "Six hundred million as one to defend China and destroy the United States invasion," but there was no stir visible in the countryside. At one village there was a straggling parade, and although it was Sunday, reduced gangs were working to double-track the railway and to irrigate the fields. To the north spread out an ever greater expanse of terraced fields yielding cotton, corn, wheat, millet and vegetables. Here and there were long untended graves and monuments, mud walls crumbling to disrepair, and villages of thatched or tiled mud houses. To the south the distant blue hills steadily approached the track. In ever increasing clarity they reared their formidable precipices until they towered above us and threatened to block our way completely.

Here at Paochi the railway line used to end. In this last bowl in the hills the late afternoon sun skimmed the encircling mountains to just touch the town. A new railway line leads south toward Chengtu through an apparently impassable wall of cliffs. We headed west, straight for a towering range of mountains that appeared to block our way absolutely until we entered the first of a long series of tunnels. At intervals we emerged from them to cross tremendous chasms, splits like giant knife-cuts in the mountains. Far below splashed the torrents that had cut these dizzying canyons; high above, the cliffs opened to reveal a slit of blue sky.

In the darkness the wheels squeaked round the curves, the carriages tilted and groaned, the compartment filled with smoke, and the engines chugged and panted. Here was railway engineering at its most exciting.

Finally we emerged on a ledge far above the Wei River and followed the sides of a great canyon, now picking up speed on a straight stretch high on the valley wall, now screaming around the sharpness of a bend in the river, now rumbling through a tunnel to gain a fresh vista of the river and the mountains.

196

Far below, the turbulent brown flood swirled in eddies around the bows of blunt-nosed barges drifting rapidly downstream or being pulled laboriously up by coolies on the tow-path. It sucked at others moored for the night in the lee of projecting cliffs. On moving vessels the crew pulled on sweeps or leaned against the tiller to keep clear of rocks, while beside the anchored boats the men sat on the banks and smoked as they watched the sunset light climbing to the mountain tops and spreading a blue-grey veil of darkness in the canyons behind.

At every cutting and tunnel entrance workers watched the trains slip by with one eye and kept the other on the cliffs above. At intervals larger gangs building retaining walls or clearing away the remnants of the latest landslides raised their yellow flags to let the train creep gently past. Clearly this part of the railway line had been built with the utmost difficulty and was only being kept open by constant effort. From Paochi it was only single track, but by next morning we had gone through 187 tunnels. Gradually the worst was over; patches of corn hung on the sloping shoulders of the hills, farmers ploughing late urged oxen across the tiny irregular fields caught on steep hillsides like patches on a Chinese suit of clothes. Small settlements appeared. Some were new homes of railway workers, built in little communes of row houses, each with a dozen one-room units, with a kitchen and a community wash-house at either end. Others were the age-old stone houses of hill-men and shepherds. These were linked by foot-paths zigzagging down between the cliffs to the main tow-paths along the river.

From courtyards among the caves and houses, the blue wood smoke curled up the valley from fires under the cooking pots. People sat round gossiping, heedless and accustomed now to the railway which had so astonishingly burst into their mountain fastness. In one village I saw three sheepdogs and in another a single one. They were the only dogs I saw in all China except for one watchdog near the Central

197

Geophysical Observatory and two—which hardly count— Pekinese in their home town, but inside the British Embassy.

For one brief stretch a wagon road wound over the brow of the hills above the river. I fondly imagined it to be the old Silk Road, but Mr. Tien, who had travelled it before the coming of the railway, told me that for the most part the Silk Road was far back in the hills away from the river gorges.

In the darkness we passed into easier and smoother going and it was possible to sleep without rolling off the seat. It was cold and I slept well from dusk to dawn with the chill night air blowing through the car and over the single thick blanket that covered all but my face.

NINETEEN

Golden city of the west

Monday, September 8th
I awoke to the white-gold brilliance of mountain sunrise and to crisper, colder air than I had felt for months. From the windows of the dining-car I watched the brightening sun search out the hollows to dispel the mist, and saw the long, dark shadows creep back from across the fields to the foot of the cliffs. We were in dry hill country, the terraced fields of yellow loess marked by well-kept walls and dotted with many graves. It was still rough enough for an occasional tunnel and I noticed that the sequence of their numbers continued up to 187 as we entered Lanchow.

Mr. Tien suddenly interrupted my preoccupation with chop-sticks to point excitedly at the window. A new concrete bridge arched over a steep defile through which the Yellow River boiled.

"That," said Mr. Tien, "is the start of the new direct line to Peking that was opened for traffic just last August. It follows the route of the Great Wall."

Between two lines of hills the valley broadened to an impressive plain with the river flooding the northern side and Kaolan Mountain standing out on the south. It had been farm land, across which industry was now spreading from

199

Lanchow. We passed an airfield and a large army depot with stores of equipment and vehicles, and many soldiers.

Over a rise in the road, a procession of men and women about two hundred strong made their way to work behind a fluttering red flag. We could hear the sound of their singing as we clattered past.

With a flurry of packing and movement in the train, we steamed into the station and our journey abruptly ended. I had reached Lanchow, the capital of Kansu Province, principal gateway to China from central Asia. Beyond it across the deserts, skirting the northern flank of the Tibetan mountains, stretched the old Silk Road to Turkestan, India, Arabia and Europe. As far back as the Ch'in dynasty in the third century B.C., Lanchow was known as the Golden City, as much for the yellow loess that has blown from the western deserts and drifted deep on its hills as for its *entrepôt* trade and its rich fields and orchards.

The old city is surrounded by a formidable wall of mud brick, thirty feet thick and nearly as high, tightly arched around a mountain spur on the south bank of the Hwang Ho (Yellow River) at a point where the valley restraining the stream broadens into a sudden plain probably a dozen miles long. The wall is breached now by many roads and will undoubtedly be taken away in time, but it is at present a picturesque fortress within which the clay walls and dusty tile roofs of the houses cling to the hills like so many inverted swallows' nests whose occupants have been disturbed by the incredible noise, the belching smokestacks, the clouds of dust, and the strange new restlessness that science has intruded into the passive old Orient.

"In 1946," said Mr. Tien, "I came to Lanchow to teach Chinese literature in the high school. At that time the railroad ended at Paochi, and I had to walk for two weeks; my baggage and books were taken in an ox cart along the Silk Road." The city he entered then, footsore and weary, had a population of 118,000 and was regarded as an *ultima*

Thule by the people of China's eastern plains. Who but a criminal, the plainsmen asked, should be sent to so desolate a spot, where the summer is hot and arid and the winter wind is cold—a place where the trees are few, the hills over-grazed and barren?

The Mr. Tien who entered Lanchow with me in the late summer of 1958 is a translator of scientific documents and papers. The city now boasts 800,000 souls. Kansu Province has tripled its old population of 10,000,000. The railway tracks no longer end at Paochi or even at Lanchow but run eight hundred miles to the north-west and are steadily approaching the Russian border. The new railway will strike directly from Moscow to Shanghai and will be much shorter than the Trans-Siberian line I had travelled. What this means depends on the map you read. On a map of China proper, Lanchow appears on the western marches. On a map of greater China—which includes the vast and formerly remote reaches of Mongolia, Sinkiang and Tibet—Lanchow marks the strategic as well as the geographical centre. The new Chinese régime is clearly planning the development of the city with that in mind.

Lanchow is lonely, as its critics lament. The mountains that tower several thousand feet above its five thousand foot elevation are forbidding. But the skies are as clear here as those over the Prairies; the hills, with some slight rearrangement, might be the foothills of the Rocky Mountains; there is enough rainfall to grow grain and support grazing, and irrigated orchards grow fruits unsurpassed in all China. Lanchow is, in short, a challenge from nature, and it is being confronted by the most enthusiastic cadres of Mao Tse-tung. Here one sees the new China at work most activily. The volume of construction under way is staggering.

Dr. Tung Chieh, an agriculturalist, the very active and busy director of the local branch of the Chinese Academy, met us with two companions at the station on the plain east of the old city wall.

We drove directly north for a mile and a half along a broad boulevard towards the Hwang Ho. Part way to the river, we stopped at an intersection with an east-west boulevard. Three lines of poplar trees divided each of the two highways into four roads, each road having at the junction a total width of ten traffic lanes. In the very middle a small circular park had been planted with trees and flowers. On the four corners facing the park were a research building of the Academy, the University of Lanchow, a new hotel, and a vacant lot.

On this formerly empty plain Dr. Tung pointed to the new city of Lanchow being built. In its present state it reminded me of Edmonton, Alberta, or Lincoln, Nebraska, when I first saw those two cities twenty-five years ago—planned in magnificent style but rather gap-toothed. There are buildings with large spaces between them that are occupied by piles of construction materials, disreputable shacks, and derelict walls and gardens. There is, however, this difference between what I saw in Lanchow and what I had seen in Edmonton and Lincoln: in Lanchow, besides scores of new buildings, scores more are in the process of erection.

The trees, planted in 1956, are already twenty feet high. Some day they will grace splendid avenues. But their present appearance is marred by the presence in every large street of ditches fifteen feet deep, in which sewers and water mains are being laid down by thousands of people working twenty-four hours a day in all parts of the city at once. The work is spurred on by the blare of gay and martial music from a hundred loud-speakers hung on lamp posts. Since the streets are largely blocked with the piles of earth from the ditches and the waiting lengths of pipe, and since the loose earth produces great clouds of dust, the cars and trucks can work their way through the mêlée of bicycles and horse-drawn carts only by slow degrees and with continuous blowing on the horn.

Having survived the bedlam as far as the hotel, I found

that establishment excellent. It was seven stories high, facing the park. I was escorted to Room 301, a two-room suite with a tiled private bathroom and a balcony overlooking the centre of Lanchow's explosive expansion.

We sat down to the two invariable introductory ceremonies of Communist China: the old custom of a welcoming cup of tea and the new fashion of making a plan for my visit. I had a chance to look at Dr. Tung, and he impressed me as a lean, active man of great energy and enthusiasm. Clad in a neatly tailored blue suit buttoned right up to the throat, his small frame carried an air of authority and decision, sharpened, no doubt, by the problems he faced in building a vast government research establishment here in the empty western part of new China.

The program of inspection tours and sight-seeing junkets was formidable, but I learned with relief that there was a good chance of my escaping a repeat performance of the scientific talk I had given in Peking. This reprieve I owed to the fact that my prospective audience were all absent from town on geological field work.

We rose from our tea and crossed the street to the new building of the Academy of Sciences. Dr. Tung himself showed us over his domain.

"The Academy's site," he said, showing us a model, "covers 863 Chinese acres." (I later learned that a Chinese *mou* is one-sixth as large as our acre.) "And the plan calls for erection of thirty-five buildings. We expect that within ten years 15,000 people will be working here and that an additional 25,000 children and relatives of the workers will also live here. At present there are 900 workers, but we expect 1,500 by the end of 1958. A recent drive to expand the staff lasted two months, but only 300 of the 500 sought could be recruited. Every district in the new China wants school graduates and is reluctant to let them leave home; so we are training relatives of the present workers to fill the vacancies. For many, this means attending classes in the morning and doing their jobs in the afternoon. The 300 new recruits in-

clude primary-school, secondary-school and college graduates."

Building of the Lanchow science centre started late in
1955. The library was finished first, in 1957. Two research
buildings and two dormitories have been completed since
then, and construction of the building to house geological
research has just begun.

I walked through the library building. It was nearly all
occupied by stacks. My private observations indicated that
the library has two wings, each five stories high; that each
story has forty bays of twelve bookcases, and that the bookcases are seven shelves high, each shelf holding from twenty-
five to forty books. This agreed perfectly with my host's statement that the library is intended to hold one million
books, and there were many indications that they were rapidly buying books to mushroom the stock they had then,
which was about 130,000 volumes. And I might interpolate
here that in every case where I was able to make a visual
check of my own or where I have had two separate sources of
information, I have found the statement made to me by Chinese scientists about quantity and date to be accurate.

"The library," said Dr. Tung, "is chiefly devoted to the
subjects in which the Academy is interested: physics, chemistry, biology, geology, geophysics, astronomy, geography,
computing, power electronics, electrical engineering, soils,
petroleum engineering and hydrobiology. Medicine is handled by another Academy whose buildings are next to Lanchow University. Agriculture is in the hands of local authorities."

The librarian told me that they subscribed to 2,600 periodicals in these specialities and as a polite gesture she produced
recent issues of a half dozen Canadian journals. The stock is
plainly at their fingertips whenever they want it. The library
building is up to date, being made of reinforced concrete
and brick, with steel shelves, air conditioned, and electrical-
ly lighted. There are alcoves for research but no reading

rooms, as borrowed books are to be taken to other Academy buildings for perusal.

From the library Dr. Tung led us to one of the two completed blocks of research laboratories and took us upstairs. There he introduced us to a project for which China clearly has great hopes. Dr. Tung Cheng, a modest scientist, has been investigating desert shrubs, expecially various species of the genus Apocynum (popularly called lopuma). It has been known for a long time that these bushes contain strong fibres in the bark; only the fact that they are very slippery has prevented their use in cloth. But Dr. Tung Cheng has found the fibres of some species to be more usable than others. He was investigating processes by which the fibres could be treated and improved, and ways of combining the fibres with wool or silk, so that they might be woven into strong fabric.

"This bush grows wild," he said, "in the driest and most alkaline deserts in the country. Its long root system enables it to survive in soil 24 per cent salt [ordinary plants die at three-tenths of 1 per cent] with only two inches of rainfall a year."

Director Tung talked hopefully of planting thirty million Chinese acres with this plant by 1962, so that the present cotton acreage can be released for growing food. I was given samples of the stuff, of which the British are reported to have bought hundred tons for experiment. To the present, the Chinese have to use existing textile machines, but now they are anxiously awaiting delivery of specially designed equipment. He gave me samples of cloth, apparently durable, of mixtures of half wool and half lopuma. I cannot tell whether their optimism is justified, but it has spurred investigation of many other native plants.

Geology is temporarily housed in an older building. Many new rock specimens in its collections show that Kansu Province is rich in minor metals. I saw celestite, wolframite and many pegmatite minerals, many base-metal ores, and also optical rock salt and oil-rich sediments.

Both the chemist in the geology division, who is the only college graduate in his laboratory, and the librarian wanted more college graduates. Incredible as it may sound, many people complained to me of the labour shortage in China, not only of skilled people, but of common labourers. So vast is the construction program now under way that I believe the shortage is probably real.

After lunch that day I was driven to the Lanchow Geophysical Observatory, which had been built especially for the IGY and was purposely secluded far from main roads and factories. The drive was therefore interesting in itself. I saw how the big boulevards of the new city joined wide roads, slashed through the old city wall. Everything was being prepared for the inauguration of trolley-bus service in October. (In a letter enclosing Communist literature that I received from Mr. Tien after I returned to Canada, he told me that he had been to Lanchow again and that the trolley-buses are now operating on the streets where we saw overhead wires being strung.) We crossed the muddy Hwang Ho and on the far bank passed three new technical schools, a home for nurses and a factory for making paper out of grass. As we turned towards the hills, we skirted fields of root vegetables grown on land strewn with river gravel to conserve moisture. The Chinese pointed to small trees freshly planted and said that to reduce the dust, grazing had been stopped on the hills around Lanchow. The hills were indeed greener there than elsewhere. Irrigation has been greatly increased too; I was told that diversion of the whole Tao River has been proposed as a means of irrigating 20,000,000 more Chinese acres. Even control of the perennially rampant Hwang Ho is not beyond this exuberant region's dreams. Our high-axled Pobeda cleared the deep mud but stalled on the final hill where the three buildings of the Geophysical Observatory perched behind a wall.

For all the loneliness of the place, I saw in the library recent numbers of perhaps twenty or thirty geophysical and instrument journals from the United States, England, France, Italy and Germany, as well as others from Russia and other parts of China. In the seismic vaults I watched the operation of nine new seismographs, mostly of Russian design and Chinese make. This would be a pretty good seismological station anywhere. The director, Mr. Chu Chang-hsung, said it was their object eventually to try to predict earthquakes.

We returned across the river to the south shore by an old steel bridge (they said there were three others near Lanchow, but I did not see them) and turned west on the new, ten-lane boulevard that runs along the river. The new buildings continued all along the route: three movie theatres, an Architectural Trust, an Electric Power Bureau, two large new hotels (one still unfinished), a broad-screen movie theatre, machine tool and implement factories, a hospital, and a whole series of new dormitories. We bumped and lurched over the road, which was all torn up. There were power lines and blaring loud-speakers everywhere. As we turned up a side road, we passed a muddy pond in which a crowd of urchins were swimming, splashing and disporting themselves. We crossed a railway line, said to be a spur leading to a coal mine, and saw a most astonishing sight. A teen-age girl was running hither and thither along the railway track in a most demented fashion, apparently beating the ground with a cloth.

"Oh, she's chasing a fly to kill it," said Mr. Tien. It was a most vivid illustration of the intensity of the effort that has been made in China to annihilate flies, bugs and sparrows. I never saw a single sparrow; I kept count of flies and saw one or more only fifteen times during nearly a month in China. In many of these instances the unfortunate fly was being pursued by a coursing Chinese. Only once did I see a swarm of flies, on a station platform near Sian, and they were buzzing around bamboo baskets, each of which held four small pigs awaiting shipment.

We turned back to the hotel along a higher terrace. Here we passed the headquarters of a Petroleum Trust, schools for veterinary science and electronics, a junior high school, a former Buddhist temple now used as a club with a large auditorium opposite to it, a mosque apparently still functioning, two more movie theatres, and masses of dormitories. The main roads were clogged with vehicles of every description, and all along the way were Chinese either digging or building. After all the excitement yesterday of "spontaneous demonstrations", we saw only one parade. The participants were all peasants, and it rather looked as if they had seized the excuse to come into town. They were very much in a holiday mood, clowning and celebrating as they marched. Seeing them prompted the director to exclaim rather sourly, "The country districts need all the high-school graduates they can get to carry out their own programs, and they hold them. Hence it is very difficult to get students for the Academy and for the universities."

Several times we passed Tibetans. I noticed the women particularly, for they seemed to be holding a convention in our hotel. In contrast to the universal plain cotton shirts and trousers of the Chinese men and women, the Tibetan women wore voluminous black skirts with very colourful orange or red blouses made brilliant by golden belts and embroidery. Their hair was braided in two pigtails that either hung down their backs below their large, flat-brimmed black hats or, if they wore no hat, each braid might be pulled up into one huge loop. They had dark, ruddy complexions.

Much more numerous were the Moslems, distinguished from other Chinese only by a white skull-cap on the men and by a black veil or hat on the women. Formerly the women wore black face veils, but these are now forbidden, and the older women did the closest thing they dared by wearing a black veil across their throats up to the chin. Younger ones pinned back their black veils on one side like a lapel or even adopted a black skull-cap similar to the men's white ones. A few people wore padded clothing that reminded one of the severe winter weather of Kansu.

Close to the hotel we got out, walked past a building pointed out to me as a Moslem restaurant and went into a three-story department store. On sale was a wide variety — cloth, clothing, shoes, utensils, tools, clocks, phonograph records, bicycles, candy, medicine, notions, musical instruments, flashlights, and limited amounts of cooked food, toys, radios, and electrical equipment, including a kerosene lamp with a transistor unit on top that by a not too complicated process converts the heat of the lamplight into enough current to run a radio. There were buyers but no queues.

Since Mr. Tien had a cold, we strolled the streets looking in small shops for the acrid Tibetan olives he wanted to take as medicine and for souvenirs for me. If I stood still a crowd of boys would gather in a moment to look me over; it was clear that no great number of foreigners come here.

From our trip around the town I concluded that in Lanchow there might be a few hundred cars and jeeps, some thousands or tens of thousands of trucks and an even larger number of bicycles. There were also several scores of ambulances that were very busy, but to what extent their business was with accidents from the vast construction effort, I do not know.

After supper I felt a desire for further exercise and persuaded Mr. Tien to go for a walk with me. Coming out of the hotel, I was as free to choose the direction to walk as I had been in the old city that afternoon. In every direction the streets were flood-lit and in the brilliant light thousands of people were digging to put in the water system. Elsewhere there were no street lights yet, and it seemed particularly black in contrast. The flood-lights shone on those shovelling and on those winding up the buckets on the windlasses. There was less traffic but trucks still rumbled by and, when they passed one another, most disconcertingly switch-

209

ed from brilliant white headlights to red dimmed lights. Fortunately, they had good turn indicators (when viewed from in front at least). It was a most ingenious rig. Inside the windscreen they flipped a large red and green lighted arrow to indicate the direction they intended to turn and were waved on by traffic police with flashlights.

On the way to the railway station we passed several more large buildings: a school of commerce, a railway workers' school, a yard for building supplies and a fifth hotel with a movie and a large store beside it. As the hotel was brilliantly lighted and no curtains were drawn, it was easy to see that it was crowded with people, all Chinese. Beside it, in the shadows, a few couples sat or strolled, holding hands and behaving in ways not unknown in the West.

There was a ditch beside the road, and at one point a whistle blew and the afternoon shift of workers swarmed up out of the ditches.

"Who are they?" I asked Mr. Tien. He stopped the ones I chose at random, and others gathered around. Through him they told me they were students in economics, that they had got up at six o'clock in the morning, had gone to classes from eight to twelve and had worked here as volunteer labour from half past two until now. Throughout the city, they said, much of the work on this "afternoon" shift was done by volunteer students from institutes and schools, housewives, and office workers. I do not know what would happen to anyone who didn't volunteer. But these students seemed not particularly tired, and they were happy to talk as long as I wanted.

I was puzzled to see so little supervision; no foremen could be seen standing around bossing the job as in Canada, but only an occasional engineer or timekeeper in a tent, or a surveyor and his assistant running levels. What keeps the people of Lanchow at work? To me there seemed to be no question that many were enthusiastic, that the force of public opinion, the need to eat, and the habit of hard work carried the rest along.

210

Having completed our interviews, Mr. Tien and I walked to the railroad station. Around it, in the open, a few hundred people were sleeping in heaps, waiting for a train. The dim light and shadows gave a wild, barbaric look to the huddled groups of sleeping children with their bundles, and the tough peasants with shaven heads, strong shoulders and chests bared to the night air while the rest of their bodies were swathed in quilts. In front of the university we came upon about sixty students and teachers, busy with a small furnace not much bigger than a barrel but flaming away brilliantly. Mr. Tien thought we were witnessing a night-school course in small-scale steel making. At the hotel the exhilarating music for the encouragement of the ditch-diggers was deafening. I shut the windows of my suite and stifled, but barely reduced its violence.

While looking around my bedroom, I found a radio set and switched it on. Idly twisting the dials past various Chinese and Russian programs, sounds of jamming, and one station broadcasting Communist news in English, I was startled to hear an American voice: "This is the Voice of America coming to you from Honolulu." I listened to familiar music until the news started and I heard the other side of the Quemoy story. I was rather disconcerted, and perhaps a shade worried about my chances of being trapped in China by the outbreak of war, as I went to bed.

Tibetan boarding-school

Tuesday, September 9th

When I awoke, I tried the radio again, but could get nothing but local programs. I also wanted a bath and a chance to wash my nylon clothes, but I was dissuaded by the cold water that came in like a splash of cocoa straight out of the Hwang Ho. I had to wait until the hot water, which was clean enough, cooled.

In the corridor opposite the elevator was a temporary exhibition of local products that the attendants gladly showed me. The chief exhibits were of fruit, furs, herbs, medicines, leather goods, wool at $3.20 a pound, and blankets at $8.00 to $16.00 each. Kansu had started manufacturing a few common articles of lacquer, glass, porcelain, and household enamel.

I was then picked up and taken across the boulevard to the University of Lanchow, where I was cordially received by Dean Lu Yun-yu and Professor Hsu Jung-on of physics. Founded in 1946, the first in Kansu Province, it now has a faculty of 700 and 2,300 students. The only subjects of instruction are mathematics, physics, chemistry, biology, geography, Chinese literature, history and economics. About 300 of the students are taking literature, smaller numbers history and economics. All the 1,850 science students take courses in physics, but only 450 of them are physics majors. The courses are being lengthened from four years to five. I found a dozen labs in the fine new brick physics building to be particularly well equipped, with plenty of instruments, a dozen air-conditioned booths for optical experiments and good facilities for growing large metal crystals and studying their magnetic and other properties. A few students were working, some of them with multilithed copies of Elmore and Sands' *Electronics*, 1949 edition, and the M.I.T. Radar School's Radar text, 1953 edition. The science in this wild and madly growing outpost of Asia is, though limited by man-power, reasonably up to date both in technology and theory.

I said to the Dean, "We met some economics students digging last night, and they told us that they did one full shift every week."

"They would not be any of our students," he replied drily. "They must have come from one of the other schools, perhaps the College of Commerce."

I got the feeling that there is a very definite hierarchy among the various universities and schools, and that while it is good for students from the crack university to have some knowledge of labour, it was considered a waste of their time and energy to do as much labour as was expected of those at lower-grade technical institutes.

Athletics were, however, considered important and the Dean proudly told me, "Recently our university has begun to compete with others in basketball and track, and at a recent athletic meeting in Sian in which teams were entered from all parts of China, we got one prize."

There was quite a contrast between the physics and geography departments. Whereas the former specialized in fundamental training, the latter did hydrogeology, practical work in land use and a study of the distribution of plants in the loess and desert regions. The students had made an excellent collection of three-dimensional models of irrigation schemes and river valleys. It is obvious that with the dearth of trained men, students get much more responsibility at an early age than they do in Western countries and that this must generate a lot of enthusiasm. Meteorology is also important in this cold and arid land, and they had a large store of all the standard weather instruments, all of excellent quality.

As we left the university we passed but did not enter the new dormitory buildings. We were then taken a little way up Kaolan Mountain to an old Buddhist temple. At the foot of the hill we passed some substantial old brick buildings. Of one they said as we drove by, "That is a Christian church; you see that the doors are open. On Sunday in Peking you could also have gone to a service."

"It is a pity that you did not mention that sooner," I said. "Can I see this church?"

"It would have to be arranged in advance and I don't know that it will be possible to fit it into our program."

They identified the building next door, adjoining the church but well decorated with large red stars of the Communist faith, as the first hospital built in Lanchow. I had no doubt that I had seen a former mission and its hospital, and that the Communists were trying to take credit for the latter, but they neither denied nor confirmed my supposition. No one was in sight at the church.

At the old Buddhist temple the Oriental buildings still stood in their pleasant park and spread out under groves of trees far up the hillside, but the interiors had been profoundly changed.

In the first main square there was a sort of exhibition attended by large numbers of farmers and officials. It was apparently a contest, with prizes, to encourage peasants in the construction of simple mechanical devices — such things as sieves, carts, water lifts, pumps and mills — by which they might increase their efficiency.

In an adjacent courtyard was an altar covered with deep red cloth and upon it a white plaster bust of Mao Tse-tung. On each side was a bowl of flowers, and around and about were paintings glorifying labour in the reconstruction of China.

Another large group of buildings had been converted to restaurants. There was no doubt that they had formerly been temples, for when I looked behind a torn paper screen, I saw huge idols glowering in the dim light as if resentful of the sacrilege.

Only on the highest terraces were a few small remnants of religion permitted. In one little chapel at the top of many steps, one or two old women, and one man said to be a monk, although in ordinary clothes, still tended Buddhas before whom tiny offerings of fruit, flowers and incense had been freshly set.

We had come for the view, and it was splendid. Before us lay the narrow plain along the valley, where we could see growing the new city of institutes, the army camp and airfield. I counted two hundred large buildings of institutes and factories, about four hundred smaller ones like rowhouses, and forty concrete chimneys, but I was not allowed to take any photographs. West of us at the defile was the old fortified city. Most of the factory area was around the corner and out of sight beyond it.

Two major groups of buildings stood out because of their gay tiled roofs.

"What are those two sets of buildings with the green roofs?" I asked.

"One is the soviet building of the Council of Workers, through whom the people under the direction of Chairman Mao govern this province, and the other is the Institute of National Minorities."

"What is the Institute? Can I see it?"

"It is doubtful if it can be arranged on such short notice, but if you would like to do so we shall try."

After lunch we drove twenty miles west through the factory district to see a new oil refinery. The whole length of the road to within a few miles of the refinery was in a state of chaos. It was the same river road on which we had driven a little way the day before. Thousands of workers were digging for sewers, widening the boulevard and constructing buildings along it. Men clad only in shorts sweated in the ditch and pigtailed girls in blue jeans, cotton print blouses and broad straw hats wound the windlasses. On the surface a few of the women were filling and carrying baskets of earth. One girl got too heavy a load and, giggling, shovelled some of the dirt out of the basket again. Perhaps some slaved because they had to, and many worked to make their living, but for some the revolution that was taking place everywhere before their eyes and within their youthful memories was a great adventure in which they were gleefully participating.

At the gateway to the oil refinery we were halted by sentries and ushered into a bare but busy office building to wait. There was little ornament besides photographs of Russian and Chinese bigwigs. Presently the manager came, rather curtly got us into cars, and drove us quickly around the refinery. Every part of it was under construction and no part was complete. I gathered that ground had been broken in 1955 and that it had been difficult to train all the welders, electricians, pipefitters and so on needed for the job. The confusion of pipes, tanks, and parts scattered about was evidence that much still had to be done.

I estimated that there were 150 large storage tanks in the several tank farms and I counted twenty groups of cracking and distillation towers, each group of from three to seven towers apiece. As they were all different, I assumed that the plant might be designed to produce lubricants and perhaps petrochemicals as well as gasoline. The plant stretched half a mile or more along the river. A mile beyond it was a very large steam generating plant providing electricity for Lanchow and the refinery.

"We expect to start production in October," said the manager. "Ultimately we will manufacture all types of petroleum products. Our oil comes from the west beyond Kansu. Have you any questions to ask?"

"I don't know anything about refineries," I said, sensing that I would not find out much and wishing to get credit for not being too generally inquisitive. "This is not my subject, but I have enjoyed seeing your plant. Thank you."

"I'm sorry if you were not interested in the refinery," said one of the local guides as we drove away, "but I have been able to arrange for you to visit the Institute of National Minorities."

It occupied one of the best sites in the centre of the city and comprised a group of fine buildings beyond which lay many basketball courts. Having parked in front of the central building, we walked up a broad flight of stone steps through a colonnaded façade and into a large assembly hall. As we came up the steps a group of young and pretty girls were intently practising folk dancing under the portico and singing as they danced. When they saw us, they ran shyly away behind the pillars.

"Those are Mongolian girls," said our host Mr. Ma, the secretary of the Institute. "The people of the national minorities love to dance and we encourage them. This is one of five such institutes, of which the central one is in Peking. Here we deal with sixteen of the fifty-one minority groups. Scholars do research in the Institute, but chiefly we provide a boarding-school giving a three- to five-year course to two thousand pupils. We train the cadres or political leaders of the future in practical ways; for example, we are training boys from Sinkiang to be workers on the new railroad being built beyond Lanchow across their province. Of course, they have never seen a railroad before and when they arrive many cannot speak Chinese, so we train them." By this time we were in a hall that was indeed an excellent and elaborate one for a school. I estimated that, with the gallery, it would hold

217

seventeen hundred. "You will notice," he continued, "that this hall is equipped with ear-phones at every seat and a switch for four-channel simultaneous translation. You can see the windows of two booths for translators on either side of the stage behind the orchestra pit." We went out of this building and across a road on which several children and two men were working. "All the roads you see have been laid by the children after school under the supervision of only two regular workers." Several hundred other children were playing basketball and chasing each other around the grounds.

The library, which we entered next, had a large reading room and extensive stacks filled with an exotic, Eastern collection of books. I expressed my amazement at the number of volumes.

"Yes," said our host, "we have two hundred thousand volumes. When you consider that we have to supply all the textbooks in sixteen languages for so many students you will understand that we have many duplicates. There are also books for research. Here are books in Mongolian." And he pointed to the writing in a vertical script similar to that I had seen on many of the strange signs about the buildings. "And these books are in Tibetan," he said, taking down bundles of loose sheets tied together between exotic, multicoloured covers and untying the cotton tapes to reveal curious Eastern printing on their pages.

We walked through the boys' dormitories. In the corridor some small urchins were squirting each other at a drinking fountain. They ran away as soon as they saw us. The bedrooms were as full as they could possibly be with single cots pushed tightly together, each covered with a brilliant Oriental counterpane. At intervals were simple communal washrooms with tiled basins and water taps.

"Here is the tuck shop," he said, and we looked into an old-fashioned shop rich with the smell of herbs and incense. "These children are accustomed to food and sweetmeats very different from ours; and to make them feel at home we provide special delicacies favoured by all the different minorities and they can buy them with their allowances." A small Tibetan boy moped shyly in a corner, still with his barbaric and colourful native costume girt about him.

"Gradually they learn Chinese and become accustomed to our ways. There is a mosque in the school for the Moslems. Of course, some of the teachers are Tibetans and Mongolians too." And indeed we saw other children and several teachers in the wild costumes of the mountain peoples.

As the gatekeeper let us out of the grounds, Mr. Tien explained that at no school would it be wise to allow children, especially those ignorant of Chinese, to run loose in the town.

The farewell dinner tendered by Dr. Tung on behalf of the Academy started at seven and lasted two hours. There were ten of us there—Dr. Tung, Mr. Tien, the discoverer of lopuma, five other scientists from the Academy, a physician and myself. We fared well on these twelve courses:

Hors d'oeuvres of hair fungus, black eggs, bean jelly, tomatoes, celery and shred.
Fish entrails or special mushrooms (I couldn't tell which)
Shrimp soufflé
Chicken in oil
Scallops with beans and greens
Gelatinous fish fins of a special variety
Phoenix tails and dragon fish in sausage with mushrooms
Silver mushroom soup with tiny scarlet egg plants
Melon in sugar candy
Stuffed and steamed pancakes
Egg and tomato soup
Honey-dew melon
Beer, wine and millet brandy

It was a delicious meal. The silver mushroom soup was

a special delicacy and quite out of this world. The scientists in the group were immensely pleased that I enjoyed their food and had taken the trouble to come to Lanchow and show an appreciation of their work. They were fascinated by colour photographs that I had of my family and summer cottage, and I took the opportunity to point out that the picture of myself on the roof had been taken when I had build an addition to the cottage with my own hands. "Although professors in Canada are well enough off to own cars and cottages," I explained, "workers get so well paid that professors can hardly afford to hire them." This was so different from their experience that I doubted that they either understood or believed me.

Afterwards, while we were sitting around having a cup of tea, Dr. Tung also started a little political campaigning. He was voluble about racial discrimination and it took me a moment or two to understand that his references to "Little Stone" concerned a place in Arkansas.

I replied, "You fail to appreciate that the difficulties there have arisen because a majority of the people of the United States and Canada, far from favouring segregation, have expressed their opposition to it; but since we do not have dictatorship, changes must be introduced gradually and will involve much public argument and unfavourable publicity."

"But the Chinese people are not well treated in America."

"I do not agree; at the present time in Canada no group has stronger representation in Parliament in proportion to its members than the Chinese Canadians. Although they form a tiny proprotion of the population they are represented by one of the 265 Members of Parliament, Mr. Jung of Vancouver."

"We have heard of Mr. Jung," he admitted.

"You must admit that he has not been discriminated against. Of course many of the Chinese who came to Canada were illiterate and had a hard time at first, but many of their descendants are now professional men. One of my collea-

220

gues, Professor Chung in physics at Toronto, is a Chinese Canadian. We also know about China from other Canadians who were born there. You probably don't realize how many children of missionaries were born and reared in China and speak Chinese as well as you do. I know several of them."

TWENTY-ONE

Hard-class sleeper

Wednesday, September 10th

The best train of the day going east from Lanchow leaves at three in the afternoon, but it does not connect with the express to Canton; so we got up and caught the morning train which has only hard-class sleepers.

Dr. Tung saw us off and pressed on me a string bag containing five delicious honey-dew melons, good apples and a bag of raisins. All supported the claim of the irrigated lands of Kansu to be the orchard of China. He hoped that I could get at least the raisins and apples back to Canada to demonstrate their excellence. He had little idea of the problems of air travel, but I liked his genuine enthusiasm and thanked him warmly for his hospitality.

All day we sat on the hard seats. They were narrow, crowded in tiers—three rows on either side of open alcoves. Fortunately the middle berth swung down by day to form a back to our seats, which were thinly upholstered in black, shining material over the boards—an innovation since the liberation, according to Mr. Tien.

The whole car, including the washroom, was more public than before and I was an object of some interest to other passengers, especially the children. As before, the porters worked hard to keep things clean, and if possible win the flag for the best car—this even extended to our porter washing the outside of the windows at one station. All the Chinese people were clean and neat. They washed themselves and their clothes and they used an instrument I had never dreamed of before—a U-shaped, narrow strip of metal, like the bent handle of a toothbrush—to scrape down the top surface of their tongues in the morning. In the thin string bags that carried their few possessions, wash basins, tooth-brushes and mugs could always be seen.

As we passed villages of mud huts all day and caves in the loess in which whole families lived, I could sense the struggle to keep clean in such surroundings, the difficulty of teaching polite manners to children living in a cave, and I marvelled at how good conditions were under the circumstances.

In the dining car, I first noticed the woman sitting facing me at the next table because she had on a pink sweater—its colour caught my eye—and then I noticed that she had a singularly vacant look on her flat and oval face, from which the black hair had been drawn severely back into a bun. She was petite and pretty, but her mind seemed far away as she gazed unseeing from the windows and unthinkingly ate rice, meat or soup indiscriminately with a metal spoon from the three bowls in front of her.

It was not until later that I noticed her blouse was modestly drawn up, and that as she dined so did the infant on her lap. It brought home to me the advantage of the Chinese style of eating with only one hand, as one does with a spoon or chopsticks, leaving the other free to gesticulate or hold the baby as the case may be. Throughout the meal the woman was nonchalance itself, even when she had to rise to let another diner pass. Chinese life today is more egalitarian than it was, and everywhere there is a direct and simple approach to the essentials.

It is interesting to notice how many Chinese use a metal

223

spoon now in place of chopsticks. Since the food is all cut up, either will serve. On the train they set a metal dessert spoon at every place instead of the conventional and rather inadequate porcelain ones still used at the Peking Hotel.

During the long day Mr. Tien entertained me with the loan of some of his private stock of reading material with which he practised his English.

I found the account of the "Second Session of the Eighth National Congress of the Communist Party of China, Peking, 1958" pretty heavy going, but I was amused to notice that they ascribe evils and perfidy and incipient collapse to their opponents, and righteousness and success to themselves, in terms that mirror those used in the West.

The United States, leader of the imperalist camp, is now in the throes of another serious economic crisis; its production has fallen off drastically and the number of unemployed increased enormously. This crisis is hitting the entire capitalist world, and has thoroughly exploded the deceptive propaganda spread since the war by bourgeois politicians and scholars, reformists and revisionists, that capitalist economy can avoid crises . . . .

The tendency to neutralism continues to grow in many capitalist countries. In Asia, Africa and Latin America, national independence movements are *forging ahead.* Though the imperialists are trying to undermine these movements by *underhand means* and by force, and though certain sections of the bourgeoisie in those nations are trying to restrict the growth of the people's forces which are most resolutely opposed to imperialism, facts have proved that they cannot hold back the historical advance of the people's national and democratic struggles.

In contrast to the situation in the imperialist camp, the socialist camp is growing stronger and more prosperous day by day. The economies of the Soviet Union, China and many other socialist countries are developing much faster than before; the living standards of their peoples are steadily improving. The unity of the socialist camp is becoming more further extended.

We may laugh at these exaggerations, but it would be

224

foolish to dismiss them as nonsense. The fact that someone had read the pamphlet before and had underlined some words and passages (indicated by the italics) and added Chinese characters in the margin illustrates that many people in China want to believe and do believe this stuff, in the same way that we in the West want to believe the news we read. It would also be foolish to think that our news of China is accurate and not coloured.

The books Mr. Tien used to practise and perfect his English proved more interesting, but they showed indirectly the same human tendency to denigrate one's rivals and flatter oneself. This description of a former Chinese diplomat, presumably of the Kuomintang party, was typical of the approach.

When one talks of diplomats one usually thinks of cynical worldlings, immaculately dressed, full of subtle cunning at the Council Board, scintillating with *bons mots* at social gatherings, irresistible to ladies, thoroughly insincere, but insincere in the grand style, charming with a seductiveness which deceives no one, brilliant as glass and just as brittle.

Finally the serious-minded Communist (and they can be very serious-minded!) considers how he should behave.

Any man of letters who strives against the natural current of changes will almost certainly be wrecked in consequence. Any book produced, no matter how well written, which can be classified with the productions of a dead school by its thought and feeling will soon be forgotten. Moreover, in your private reading it is very, very essential to read in modern directions.

Perhaps I flaunted my lack of conviction too cheerfully, for after listening to the Chinese news Mr. Tien turned to me and warned,

"The broadcast says that the people are volunteering to work harder and to fight if necessary to preserve our country from invasion and to liberate our people in Taiwan.

"Speaking objectively, however, I must admit that no one is doing more to help our cause than Mr. Doo-les. By stirring up our indignation and mobilizing our people, he is unifying our country behind the Government."

Think what you like of these views, they are those being drummed into hundreds of millions of people — people who like ourselves want peace, and like ourselves are fearful of strangers and want to bolster up their own self-esteem.

TWENTY-TWO

Chinese thé-dansant

Thursday, September 11th
All day the train ran slowly down the Yellow River valley
by the same route we had come. We passed Loyang, and
finally at 4 p.m. reached the place shown on the map as
Chenghsien, but called Changchow by Mr. Tien, the junction with the main line from Peking to Canton. We got off
and the hard-class train continued on to Shanghai.

An Intourist man drove us to a small modern hotel for
a bath and rest, and I noticed with some interest that the
somewhat primitive bathroom had only the sign written in
chalk on the door, "Hombres." I pondered over the strength
of Communist influence in South America, but the only
Caucasian I saw was a Russian.

At 5:30 p.m. we went for half an hour's drive around the
town; they told me it had been largely destroyed by the
Japanese, and it was clearly being rebuilt as somewhat of a
show piece. Conditions in the old part of town were still
crowded and primitive. The main roads had been widened
and were concrete, but they had open drains and water was
obtained from big pumps at intervals along the streets. On
one corner were large, new buildings for a post office, a big
movie theatre, and a department store.

227

We then drove around the new cultural district to the north-east, going by the main boulevards—wide paved streets divided or lined with rows of trees planted in 1956. Along them were a great number of institutes and dormitories; my hosts identified a hospital, a medical school, schools of architecture and commerce, an Institute of Town Planning, schools of forestry, hydraulic engineering, and agriculture. There was the headquarters of the Yellow River Conservancy Board, two new, large hotels, many schools, movies, shops, and so many dormitories that they told me that all the inhabitants of the new part of town lived in them as well as a considerable portion of those in the old town. For the first time, in this city of a million people, I saw posters boasting of the improvement in housing.

The contrast to Lanchow was remarkable. This was a settled place and more peaceful, with fewer vehicles and more bicycles. "This city," said Mr. Tien, "used to be at the geographical centre of China. That has now moved to Lanchow."

While having dinner in the hotel, I heard music which sounded as though it came from a jazz band in another room; and when we left the dining-room, we passed a hall in which a hundred or so young Chinese in Western clothes were sedately doing the fox-trot to the music of a band that boasted a clarinet, two saxophones, trap drums, a large horn and a trombone.

Mr. Tien, the Intourist man, and I looked in. They offered to introduce me and invited me to join in the dancing. However, I left that to what I assumed to be the young *élite* of the local Communist party. Far from being a bacchanale, it reminded me of the stilted *thé-dansants* popular for young people in the nineteen-twenties.

At one end of this room, and again in the library, there was a buffet or table on which was a white plaster bust of Chairman Mao, with flowers on either side and a mirror behind. These arrangements seemed very close to being altars and Mao to being a demi-god—the only god modern China has.

It is an interesting point for debate whether the worship of Mao will grow into a new religion or whether the attenuation of revolutionary fervour and the inculcation of learning will inhibit this. Although the Chinese are notorious for being a people not strongly influenced by religion, many of them clearly see in Chairman Mao the restorer of Chinese omnipotence, the leader who will unify all of the Chinese peoples, and the heir apparent, in fact if not in form, to the Dragon Throne. For many Chinese he has been the executioner, the cause of ruin, the destroyer of happiness and ravager of all that life held worth while. For more he remains the successor to the absolute power, the temporal glory and the filial respect paid for three millennia by the humble Chinese to the Sons of Heaven.

Through dense crowds of Chinese pouring out of the movies and walking or bicycling home in the dark, we drove to the station and boarded a good sleeper on the Peking-Canton Express.

TWENTY-THREE

Brooks too broad for leaping

Friday, September 12th

This was a most superior train, or so it seemed after two days in the hard-class sleeper. Mr. Tien and I shared our large, soft compartment with a flourishing aspidistra plant on the table beside a purple-shaded lamp. It got increasingly hot on the plains, and this brought out a strong smell of disinfectant. I missed the mountain air.

All morning we hung out of windows and doors, or during the stops stood on old station platforms, as the train inched forward through the triple metropolis of Hankow, Hanyang and Wuchang, now grown together and collectively known as Wuhan, although the city is still divided into three by the great Han and Yangtze Rivers at their junction. Two great new bridges thrown across these wide rivers have made this unification possible.

For Mr. Tien the crossing of the Yangtze by one of these bridges was clearly the high point of the trip. The precautions associated with this crossing were no doubt the chief cause of our delay. The bridge over the Han was impressive enough, but the Yangtze bridge—the dream of the Chinese for centuries and the greatest single engineering feat of the new régime—was indeed quite exciting. It is a double-decked, steel bridge with six lanes for vehicles above and two tracks in a tunnel of girders, the whole supported for a mile or more on a great series of piers across the mighty river. After our long wait and a big build-up by all the animated Chinese, we rolled onto it at last, watched by dozens of excited sentries with tommy guns at the alert. They appeared to fully expect some Nationalist spy to throw a bomb out of our train at any time, and they were taking no chances. Although we were higher than mast height, we could see no ocean vessels in the wide channel but looked down on river boats and junks floating far below.

The cities were old, busy and crowded. There were many new factories, but no transformation of whole communities in the manner of Lanchow and Chenghsien.

The crowd at lunch were much gayer, noisier and more sophisticated than the passengers had been up-country. No doubt they were elated at crossing the bridge, and a larger proportion on this luxury train were senior persons and ate in the diner.

During the afternoon we passed steaming rice fields, a huge lake or two, and bare red hills newly planted with young trees. Mr. Tien found a friend of his, another interpreter, who looked after five Russians on the train. He came and chatted with us for a while. I discovered that he was a keen linguist. He was fluent, so he said, in Japanese as well as English and Russian, and he was studying German. His only knowledge of Canada was that it was cold and was the home of Bethune of Canada, about whom he knew much more than I.

"Dr. Bethune was the ideal internationalist. If everyone was like him, there would be world peace. That is why a hospital and a marble memorial have been erected in his honour. Why is there a United States embargo on literary books, so that we cannot read his life in English?"

At one station there were large, wooden tanks full of water and small fish. A lot of men were unloading buckets full of more live fish, and other men were pedaling pumps to aerate the tanks. I was very curious.

"Those fish," I was told, "are being moved from Tingling Lake, which is so big that you can't see across it, to stock the Ming Tomb reservoir." We passed along the shore of the lake later in the afternoon and, in my ignorance, I was astonished to see so large a lake in China.

On the platforms the food for sale was better and more varied—individual bowls of cooked rice complete with chop-sticks, over which a bowl of hot stew was freshly poured, "sandwiches" of stuffed and steamed pancakes, wheat dumplings and noodles in bowls.

"The south of China is richer and has better food. If people wear simple clothes and no shoes, it is only because of the heat. The southern women also are very strong and can do the hardest kinds of work."

On another platform a dozen children in uniform, each wearing his name on a metal pin like the number badges on the train porters, showed a great interest in me. It appeared that they had never seen a Caucasian or at any rate a non-Russian one. None of them knew a word of English, but they were in middle school and boasted of the small iron furnace they had been building after school.

Back in the car, Mr. Tien enlarged upon his political views.

"The Confucian scholars were hypocrites," he said. "While they preached morality, they kept concubines; while they extolled virtue, they exploited their tenants. They were landlords, and they owned seventy or eighty per cent of China. The people had only what remained. It is not astonishing, is it, that the starving people became bandits? Since the liberation the landlords have been killed and the land has been given back to the people, who are now organizing communes to work it more efficiently. There are no bandits now."

A revealing remark followed.

"Now, since everyone knows everyone else's business, how can anyone become a bandit any more? In any case the people live in peace together and try to help one another. Why cannot everyone live in peace together? Why does Mr. Doo-les prevent twenty-four American journalists from coming to China? It would be only fair to let twenty-four Chinese journalists to go to United States in exchange, would it not? Why does he prevent it? The United States invasion of China is wicked. No one supports the American imperialist aggressors except the British and the South Koreans. Did you know that Mr. Sidney Smith said yesterday that Canada would not support the United States in a war against China?"

"As you very well know, Mr. Tien, I have no means of knowing what goes on. You are my only source of information and without more of the facts it is impossible for me to argue with you. There are always two sides to an argument. How can I discuss what Canada is doing with so little information?"

"Why do the English support the American warmongers? The British are nothing but pirates and shopkeepers. It was the British Opium War that started the revolution."

"Good for them!" I said. "At least you can't quarrel with the result." I knew that his bark was worse than his bite, and I soon had proof of his obliging nature.

Investigating a commotion in the corridor, I found two perspiring and irascible Englishmen shouting at our bewildered porter, who like so many foreigners couldn't understand the plain King's English even when it was delivered at the top of the speaker's lungs. Playing the boy scout, I offered my interpreter's services.

"Thank you very much. If it hadn't been for the incompetence of that chicken-hearted pilot we should have been in Canton an hour ago enjoying a good bath instead of being dumped in this god-awful place with no arrangements, no Intourist and no one who can understand a word of English."

233

Mr. Tien soon had them straightened out, and I had the temerity to visit their compartment to ask if everything was now all right. They were Queen's Messengers carrying dispatches from Peking.

"These Chinese pilots! Won't fly at night, and as soon as they see a thunderstorm ten miles away they go to ground. We could have flown around it perfectly well."

The storm of their wrath was dying down, but they were still rumbling gently when I ventured to ask about food and suggested that there was a dining-car forward.

"Wouldn't touch their ruddy stuff. It gives me dysentery right away. A bowl of soup perhaps, but it's too greasy for me. Chicken! Probably was hatched in King James's time. Couldn't cut it with an axe."

However, Mr. Tien quietly fetched a waiter and after further altercations they ordered an omelet, cold water, and sliced tomatoes—the last two a most lethal diet to my way of thinking—and they ate these with biscuits and instant coffee that they produced from their luggage.

When all was calm and Mr. Tien and I were safely back in our compartment, he remarked to me musingly, "We Chinese say of the British in China, 'When in Rome, do as you do at home.' "

TWENTY-FOUR

Chinese Marseille

Saturday, September 13th

We awoke early and saw the gangs of workers plodding in single file across the fields to work. Wearing straw raincoats beneath their heavy coolie hats they moved like rows of animated toadstools. It was clearing after the rain and in the grey and steamy dawn men and women stood knee deep in muddy water to plant rice or to goad their water buffalo. Scattered among the unending rice fields were bananas and bamboo, and higher up the hillsides new terraces were being formed for cultivation. Higher still on the steeper slopes, gangs were busy setting out saplings.

"It is part of a great program of reforesting southern China with pine trees," said Mr. Tien. "The trees have not been keeping up with the people."

In the next compartment three Japanese gentlemen in prim white suits sat, their legs neatly crossed and tucked up beneath them, on straw mats spread out on the seats of their compartment. At one of the innumerable halts our train made that morning, one of the Englishmen pulled his head back from looking out of the corridor window and grumpily turned on me: "When does the last train leave Canton for Hong Kong? If the blighters don't get a move on we'll miss that connection. Could have made up that four hours last night. What are they doing? We've been sitting around this bloody place for half an hour and nobody's doing a hand's turn." Certainly the service was much delayed all along the Peking-Canton line but whether it was due to floods or some other cause I never ascertained.

About noon we got to Canton and were met by Director Hueng of the local office of the Academy, who took us to the ten-story Ichung Hotel, obviously an old landmark, on the waterfront just west of the Pearl River Bridge. They gave me a good corner suite on the top floor looking north over the crowded city and east over the noisy and pungent river. Moored along either bank were thousands of houseboats jammed like logs in a boom, and every imaginably type of vessel floated with the stream or darted across the river like a pond-skater.

In the restaurant there was a great assortment of nationalities—Europeans, south-east Asians, Australians, Russians and another couple of Canadians. I had a brief chat in the elevator with an East German who told me that he was an expert in botany and agriculture. I was astonished to hear Mr. Tien using English to order our lunch. It was a bizarre but sensible expedient, since the Cantonese waiter could not understand Mr. Tien's Mandarin, but like most servants in the principal hotels of this cosmopolitan city, understood English. It is true that Mr. Tien could have written the order, since the Chinese written language is universally the same, but it was simpler and quicker for him to use their common language.

We drove east of the city, past the headquarters of a railway and its dormitories, to visit four new universities—Chinan University for Overseas Chinese, a teachers' college, and schools of engineering and agriculture. Unfortunately it had come on to rain in torrents, and so we only drove around the extensive campus and new buildings of each in turn without getting out of the car.

236

On the way back to town it cleared and we stopped to walk around three memorials. The first was old, erected long ago in memory of the martyrs of the 1911 revolution against the emperors, with inscriptions in English indicating that donations had been received from societies of overseas Chinese in a hundred cities all over the world. The next was a new park, scarcely complete, dedicated to the Canton commune of 1925, apparently the seed from which the new government had sprouted. The third was a splendid hall, apparently old, commemorating Sun Yat-sen, the leader of the 1911 revolution. Near it was a great new science exhibition and museum building that we did not enter. Behind it on a hill was an ancient five-storied pagoda, ninety-two feet tall.

"This is an historic shrine," said Mr. Tien, "but it was damaged by the British. Now it is a museum devoted to the history of the revolution from its inception in the Opium War until the liberation."

A series of halls and stairways, with exhibits on each floor, led up the large tower. The ground floor was devoted to a large statue and tablets in praise of Mao; the second and third floors to displays of pottery, bronzes and pictures illustrating the rise of Chinese civilization from Peking Man onwards; the fourth and fifth floors to photographs and mementos of the revolutionary leaders and of the struggles of the Chinese people against the foreigner from the Opium War in 1840 to the present. The internal struggle against the tyranny of the emperors seemed to be played down. Behind the hill was a large swimming pool full of young people (Mr. Tien admitted it was not new), and a gymnasium in which some show seemed to be in progress, for about a hundred cars were parked outside. A larger proportion of the cars in Canton were British than was the case up-country, but the city buses were either Chevrolets, ancient Dodges, or the new Chinese make.

At the corner of the Pearl River Bridge is a large, handsome new hotel for overseas Chinese visitors and next door to it a five-story exhibition building belonging to the Export Trust. It was barely finished and the management were busy arranging a new fall showing, but we were made welcome and escorted all through the building.

The goods displayed covered a wide variety of needs, but without frills and without the competing lines of a free-enterprise economy. It was like a mail-order catalogue with few choices, designed to meet the essential requirements of modern Western civilization in as economical a fashion as possible.

The exhibits for export were all neatly arranged on stands or in showcases, and all were said to be produced in China. I noticed one bus, a line of machine tools, small electric motors and tools, bicycles, a few light steel girders, rails, iron pipe up to six inches in diameter, other common metals in various shapes, raw materials such as sulphur, talc and barite, food galore, lacquer, good furniture and chests, carriages, toys, rugs, cloth and clothing (much of it beautifully embroidered), household utensils, dyes, stationery, surveying and drafting instruments, microscopes, radios, electronic components, telephones, electric meters, chemicals, glassware, 35 mm motion-picture projectors, cameras and film, typewriters, waxes, rope, leather and straw goods, tires ("Rubber is the only raw material imported into China"), paints, brushes, tools and shoes. In the best style of Western salesmanship we were presented with lithographed catalogues and copies of a monthly sales journal.

It was in this building at half past three that a bell rang, and twenty scrub-women stopped washing the floor and started a tirade. The younger and more enthusiastic stood up among their mops and pails and did the rhythmic physical jerks approved by the Chinese authorities in lieu of a coffee break, meanwhile keeping up a running fire of abuse at the tired, older women still on their knees who obviously considered the whole performance silly.

The tour the authorities planned was apparently finished, and so once outside I told Mr. Tien I wanted a particular type of Chinese doll to add to my wife's collection as a souvenir. Willing as ever, he set out to find one for me and we drove hither and thither around Canton in search. Since there were masses of shops and plenty of things to buy I found this an interesting experience and a good way to see more of the town. I refused, therefore, to be satisfied and persisted in dragging Mr. Tien all around the centre of Canton. This city, although no doubt much cleaner than in former days, was not as spotlessly tidy as Peking. It was more crowded, more lush, more varied and richer than the north. Some women shopped in neat and almost opulent black silk tunics and trousers, others wore the slit-skirted Chinese dresses and still others, poor and bare-footed, heaved and pushed in groups of two or three at great hand-carts laden with a ton or more of top-heavy merchandise. Some even carried babies strapped to their backs while they struggled, which pointed up the hardships of life in China.

A lot of changes were going on in the city. Many shop fronts were being renovated. Some whole streets were being rebuilt and widened. Complete stopping of traffic had been accepted, but in order that the shops would not have to close during the reconstruction, they had been moved out into temporary sheds of poles and matting erected along the centre of the streets. It was queer indeed to see these rows of busy shacks, while on either side of the road the buildings had been torn out as effectively as if they had been bombed.

Much more widespread was a ripping out of metal grills, shutters, and stairways to be replaced by wood. Mr. Tien seemed to imply that this was an improvement, but it looked like a vigorous scrap metal drive to me. At one place he pointed excitedly and said, "There they are constructing a community dining-room. Soon everyone will be more efficiently fed."

After supper we continued our reconnaissance on foot,

covered by the dark tropical sky star-lit above the river, surrounded by the dense week-end crowds of deferential Chinese, and immersed in the smells of the East as thick as herb soup. Sweating through the humid night, I felt as soggy and alone as a single grain of barley among the rice that had sunk to the bottom of the bowl.

TWENTY-FIVE

East train to the West

Sunday, September 14th

At 8:30 we were put on board a comfortable day train by Director Hueng and a charming young woman official, with many expressions of hope for a safe return and that my trip through China had been up to my expectations. They were sorry that shortage of time and inclement weather prevented me from seeing more of their city and its surroundings, but they trusted that my visit had nevertheless been of some interest and profit to me.

Continuing in this vein, Mr. Tien apologized to me that the season was not propitious for giving me a meal of that acme of southern cooking—flesh of poisonous snake. It is apparently poisonous in more ways than one, for he explained, "We only give this delicacy to visitors in winter. In summer there is no guarantee that it is safe. Some Chinese used to regard it as so delicious that they ate it during all seasons of the year and a number died in agony every summer."

Mr. Tien showed his appreciation of my consuming interest in Chinese food by presenting me with a cook-book of fifty recipes in English which he said my wife might like to prepare.

I seized the chance to ask about bird's nest soup.

241

"It is very scarce, for it can only be obtained from the nests of one type of bird living high in the mountains. It costs $6.00 or $8.00 an ounce. The silver mushrooms for the soup at the Academia Sinica banquet in Lanchow only grow on the borders of Yunnan and Szechwan, and they cost about $3.00 an ounce. Although they swell up greatly in water, it would have required at least two ounces for the ten people at the dinner. Formerly both these delicacies were reserved as tributes to the emperors."

In three hours we reached Shumchin, the terminus of the Chinese railway, and got out at a comfortable station. We were given tea in one of the several waiting-rooms and had our lunch while officials examined my papers, which they had taken from me as we had stepped onto the platform.

In due course they returned with my passport, waived any examinations of my baggage, and told a porter to bring my bags. At the end of the platform a creek flowed under a steel railway bridge over which the line had once continued to Hong Kong. Only a barbed-wire fence and one or two casual sentries marked the border. I shook hands with the faithful Mr. Tien and thanked him warmly for looking after me so well. I asked him to convey to the Academy my appreciation of their hospitality and wished him a safe trip back to Peking next day. He had already written in my diary these heartfelt expressions of his sentiments:

> With close co-operation
> in the fields of science
> and culture between
> our two countries.
> Wish your visiting
> strengthen the friendship
> between China and Canada,
> put sciences at the
> disposal of mankind.
> Long live peaceful co-existence.
> Long live Peace. Amen!

242

Half way over the bridge I turned. We waved and went our respective ways.

Across the bridge I entered Lo Wu station through a series of customs offices. Several Chinese of the Hong Kong Police stood about looking very well drilled and smart in the starched khaki shorts, peaked caps, black belts and shoulder straps of their military uniforms.

No one paid the least attention to me except one British officer who, sitting back in an easy chair, waved a salute at me as I passed the door of his office. They had evidently been shown my papers by the efficient Chinese.

On the platform a representative of Chinese Intourist in a white uniform told me that the city was still twenty miles away. He asked if I wished to travel first class, gave me a ticket and showed me to a seat, telling me that he would see that my bags were put on the train.

He would have made an excellent guide for a Cook's Tour and I daresay that was how he had got his training. A news agent appeared with recent British newspapers, English and American cigarettes and offers of whisky, a Collins, Dutch lager or Hong Kong beer. With a sudden feeling of relief I turned and talked to my companions—an Australian wool merchant, and buyers and salesmen from Switzerland, Austria, Finland, Colombia and other parts of the world. Some were interested only in catching up with the news of the outside world, about which all of us were hazy, but others spoke of their experiences. One said, "I had great difficulty. The men who used to run the business have all disappeared—been shot I suppose. The new managers neither understand the market nor dare to make up their minds. In order to disguise their incompetence they gave me a long political talk every morning for weeks. It was very difficult to get any decisions at all." Others had been more successful.

We ran down valleys between rough, green hills, passing patches of farm land and pleasant sunlit coves full of yachts, fishing boats, and junks. Apartments and factories appeared and increased in number until they covered the hills around Kowloon station, where I was greeted by two Canadian medical missionaries, Dr. and Mrs. Kilborn; several newspaper reporters, and the Chinese Intourist man. It gave me a great feeling of relaxation and pleasure to be so warmly received by the Kilborns and to be taken on the ferry with its smell of the sea to their home overlooking the gay and busy harbour.

Broad thoughts from at home

It would be easy to end this account of my journey through
China here at Hong Kong, in university surroundings
among charming acquaintances with a familiar cast of mind,
but I might leave the impression that I was not aware that
China poses many serious problems for us as well as for the
Chinese. What is more, I would not have explained why my
view of China was more favourable than I had expected or
than the reader would expect. Indeed, anyone who has read
the description of China common in the American press for
the past several years would expect me to have seen little but
brutality and misery.

For anyone who has been in China for only a few weeks
it would be presumptuous folly to chart a course dogmati-
cally; but one may, at least, mark some of the shoals. I feel I
must mention some of the problems even if I cannot presume to offer simple panaceas for them, because a proper
understanding of what is happening in China today is essential to our future safety as well as that of the Chinese.

First I shall point to the dark side of the picture, of which
the West is well informed and of which I saw little, and
then I shall summarize some of the pleasant things, which
are also true and which I did see. From a consideration of

245

these two faces of China we can, perhaps, draw a balanced picture of conditions there. We cannot frame a sensible policy until we have such a picture.

Of those matters of which it most shocks one to read, and upon which the imagination dwells with horror, I saw little. Some had happened several years ago and were now passed, others were intangible, and others were kept from me. However tragic a disaster may be to those who have experienced its horrors, it will probably mean little to those who later only view the site. Cold ashes do not harrow the mind like a conflagration.

When I came to China the tyranny of the Empire, the bloody civil war, and the Japanese invasion were alike over and done. The expulsion of the Christian missionaries and of Western business men had taken place years before and they had vanished so completely that, to the casual observer, they might never have been there. The revolution in Chinese family life, philosophy and religion was impossible for me to measure, for I entered no Chinese homes, viewed no Chinese hearts, and could make no comparison with the past. The Communist Government meant little to me, for I was not dealing with politics but with science.

The acceptance of Russian doctrine and help was inconspicuous. Even the repudiation of American ideals, generosity and military aid did not disturb me as it would have hurt an American. I was sympathetic, but my taxes had not paid for the U.S. Army equipment, abandoned by the Nationalists, which I saw used in Chinese reconstruction. I could not read or interpret the hate campaigns that I believed to be swirling about me; in any case, they were not directed at Canada (which I never saw mentioned), and certainly not at me. That which one does not see makes little impact, and not being a student of China I was not fully aware of the changes and the loss.

On the other hand there were many things that I did see and that did impress me very favourably. Many had always been there, though I found the mountains of western China even higher and grander than I had imagined, the plains more extensive and more fabulously fertile, Chinese architecture and antiquities more unusual and fascinating. The Chinese people and their art, their language, their delectable food, their turn of thought and expression have always intrigued and delighted visitors. These have not changed.

Some good things are new. Everywhere are rapidly growing schools and colleges. The material resources of the country are being developed at a break-neck speed. Railways, power lines, factories, offices and dormitories are springing up all over. The Nationalists could very well argue that they would have done the same, but the Communists have the enormous advantage of having control of all China, except for one small island, and they are the ones who are doing it.

The scientists among whom I travelled had a genuine interest in international ideas, but in the country as a whole I sensed an overwhelming surge of national pride. The foreigners, hated and envied, had been thrown out; China had discovered that it could advance on its own and restore its own celestial prestige.

Above all, the Chinese did not give the appearance of living in a slave state. There is certainly a strong government, but the uppermost impression received was not of a terror imposed by the police but rather of a genuine and tremendous impulse of patriotic fervour. This fervour has undoubtedly seized vast numbers of the youth of the country, who are working for China and for themselves with a zeal and energy which I have seen matched only once before in my life. This was in England, in the warm summer days following Dunkirk, when invasion threatened the coast, dog-fights flared in the sky, factories filled to the shout of "Go to it!" and through the black-out of the scented evenings one could mark the pubs in Kent by the chorused jingoistic jingle, "There'll always be an England."

Anyone who knew China in the past can list such a multitude of wrongs and agonies that her people have suffered as would break the heart of a thousand Jobs, and all may be true. On the other hand, anyone who goes to China today sees before him a myriad of new accomplishments and a far-sighted planning for the future. It is of vital importance that we should assess what is happening as accurately and dispassionately as we can, because what happens in China is going to affect us—probably with a smash of tremendous impact—and that soon.

It will affect us because the world has so shrunk that disturbances in one part react upon the rest. It will affect us because modern weapons are overwhelmingly quick and destructive. It will affect us because China is growing so fast in population and strength that it is not likely to stay contained for long. The population of China, which is some 650,000,000 people, is increasing at the rate of about 17,000,000 people a year (a new Toronto or Washington every month, a new Canada every year and a new United States every decade). By about 1980 the population may reach a billion, and that is approximately all the people China can hold.

Within twenty years, therefore, one of four things is likely to happen: the Chinese must introduce some form of effective birth control, they must produce food by some radically more efficient means, they must suffer some prodigious natural calamities, or they must expand. At the present time, although both the first two are possible, neither of them appears likely. Both the others would certainly involve their neighbours and perhaps all of mankind.

The Chinese are clearly more aware of this sword of Damocles than anyone else, and the present Government is taking measures that will enable it to face whichever of these eventualities materializes. It is making the whole population literate and is indoctrinating them. It is industrializing the economy. It is moulding and organizing the entire country into an immensely powerful tool in its hands. The problem is terrible for China, but it is likewise a terrifying prospect for us, and the discovery of a satisfactory solution, if one can be found, is as important to us as to the Chinese.

The world is much too small and China much too big for us in the West to suppose that we can continue much longer to ignore what is happening to the hardest-working quarter of the world's people. It is very easy to indulge ourselves in a hate campaign against Chinese Communism, but the more one fears Communist China the more important it is to help the Chinese to find some solution. To seek a solution it is vital that we understand clearly what is going on in China.

In the first place, one must admit that many views are possible. As Sir Josiah Stamp once said: "If two men see a dog and one of them thinks it is a wolf, their reactions will be different."

There is no question but that in the revolution millions of people were executed, most without trial and many without cause. Other tens of millions were ruined and saw their relatives, their homes, their religion and all the dear, familiar things swept away. They will never forget, never forgive. But how do we know that there are not among the hundreds of millions of Chinese a vaster number who rejoice?

The average Westerner who went to China, whether scholar, diplomat, or business man, came in contact with the upper class of Chinese. He did not know the coolies who worked for a pittance. Today it is the coolies and the peasants who have seized control. We hear the cries of our injured friends but, perhaps, the feelings of present-day China are more accurately represented by the war-whoops of the risen peasants. If one thinks how they had suffered for centuries, can one blame them? That view of China is not one that the West likes to accept, but it may well be true.

If we are to achieve an objective appraisal, we must recognize the existence of bias in our thinking about China.

American feelings still smart from the repudiation they have suffered. Having poured treasure, missionaries, arms, advisers and affection into China for generations in a way that was more altruistic than the behaviour of the Russians or other European nations, they are justifiably hurt at the paradox that it is the Russians who have been accepted as mentors while the Americans have been made the chief target of vicious and unwarranted hate campaigns.

Americans, sensitive to these wrongs, have had no chance to see China for nearly ten years, and their resentment has grown, unchecked by any contact with realities. On the other hand, people who go to China are attracted by the positive things they see. It seems that the mind tends to magnify unseen evils. This would explain the disparate views of those who merely read of present-day China and those who have seen for themselves. I have talked to about sixteen Canadians who have been in China in the past two years and I have read views of nearly as many others. They represent all shades of opinion, from ardent left-wing writers to conservative newpaper editors and bank presidents, but without exception all found much that was praiseworthy in China. One of them, Walter L. Gordon, a most respected chartered accountant, business man, and adviser to the Canadian Government, has published in many Canadian newspapers this description of a May Day parade:

The May Day parade was a spectacular affair and wonderfully well organized. The marchers and performers are carefully chosen, in some cases by their fellow-workers, and obviously they must receive considerable training before the show. It is considered a great honour to participate. The parade marches past the Tien-an Men (the Gate of Heavenly Peace), a long, high and most imposing-looking structure which dominates the People's Square and is the gateway to the Forbidden City. An open gallery runs the full length of the Tien-an Men about sixty feet above the ground. It was upon this that the nation's leaders assembled, with Mao Tse-tung in the centre. We had excellent seats, or rather standing room, just below and slightly to one side of the centre of the main gateway. From this vantage point, we had a clear view of the leaders just above and behind us, and of the parade as it passed in front.

Mao Tse-tung, whose bust or photograph appears in every public building, in every meeting hall and, it seemed, in the majority of the individual living quarters we visited, is a bulky, strong-looking man of sixty-six. His followers believe he will be remembered as perhaps the greatest man in history, certainly the greatest in the history of China which, after all, in their minds is synonymous. He stood for two hours and a half reviewing the parade and at the end looked much less tired than we were. Reports that he is in failing health would seem to be exaggerated.

The march past was scheduled to take two hours and a half. It began sharply on the stroke of 10 a.m. and the last marcher passed the reviewing stand at 12.30 exactly, which is an indication of how well these people organize. The marchers included peasants with exhibits depicting cotton, wheat, pigs, sheep, etc.; representatives of the national minorities, all of whom were very colourful—Tibetans, Mongolians, Moslems, people from Sinkiang, etc.; factory workers from various plants and industries; athletes, students, dancers, acrobats, some of whom were doing the giant swing on a horizontal bar time after time; football players; floats with ballet dancers; a women's army group; a few aviators; and more dragon dancers; dragon boats, Chinese lions; the Peking opera, etc. Periodically they let off balloons, to which were attached in some cases a big papier-mâché dragon or an enormous horse depicting The Great Leap Forward of 1958. Everyone seemed to be having fun.

There were a few traffic policemen and a few unarmed soldiers lining the route in front of the reviewing stand who stood at strict attention for more than three hours without moving. But only a few soldiers appeared among the marchers and there were no guns, tanks or aircraft. We asked about this and were told these could be seen in the parade to celebrate China's National Day next October 1st, which will be the tenth anniversary of "Liberation". But that the Armed Forces are not featured on May Day, which is the traditional workers' holiday and celebrated as such. It was surprising to us, therefore, on returning home to read the following paragraph in an account of the May

Day parade which appeared in one of the weekly news magazines that is so widely circulated through the Western world:

"But behind the 'spontaneous' revelers this May Day in Peking, came squadrons of Russian-made tanks of the 'People's Liberation Army'. Forktailed MIG-17's dipped overhead, while the infantry, 110 abreast, marched in grim determination."

Undoubtedly the Chinese do have Russian tanks and planes, but they did not take part in the parade in Peking on May 1, 1959. We happened to be there.

The present climate of American opinion creates its own literature which feeds the fire of its fear and hate. Any thought that the news of China published in the United States gives a fair picture was banished from my mind after speaking to a group of business men in New York. I referred, without mentioning any names, to an account published in a well-known American weekly of a function I attended in Peking, of which I have deliberately omitted mention in this story. I suggested that the published account was biased. Afterwards, the publisher of the magazine, who had by chance been in the audience, came up to me, said that he understood the reference, sympathized with my views, but explained his own stand by saying: "We can't move faster than our public."

Just at a time when the exchanges between the Soviets and Western powers are increasing and their diplomatic relations are slightly improving, blind hatred and fear is producing a situation of the utmost rancour and mutual antagonism between the United States and China. Each side hates and fears the other. The Americans are entitled to remember with resentment the Communist uprising and the attack on Korea, but the Chinese likewise resent foreign ships and planes off their coast. Each reviles the other, and neither wishes to draw back from positions that are obviously dangerous to both. The situation is so tense that no simple action can quickly ease it. A solution can only be effected over a period, with conciliatory moves on both sides. Perhaps the situation in China can be more clearly understood if we compare it with other examples from history.

If hatred of the luxury and excesses of the French court under Louis XIV and his successors led to the bloodshed of the French Revolution, then it was exasperation at the utter fecklessness and bumbling of the Manchu emperors that brought on the Chinese Revolution. If at the end of that revolution there was bloodshed, then we do well to remember the blood that was shed in France.

The American colonies, less than two hundred years after their foundation, rose to free themselves from a despotism that was, in point of fact, not severe. How much more readily then should the Americans understand the causes of the Chinese Revolution! Indeed most Americans welcomed the start of the revolution in 1911, and hailed Sun Yat-sen and Chiang Kai-shek as leaders. As Burke hailed the French Revolution at its inception and lived to deplore its monstrous development, so the United States has recoiled from this horrible child to which it stood godfather.

We in Canada have a particular reason to sympathize with the Nationalists exiled in Formosa, and we can also very well understand that history forgets the losers. It was the losers in the American Revolution — the Tories — who were forced to migrate to Canada and so became the first large body of English-speaking Canadians. As a Canadian, I am particularly aware of how foolish it would be to continue to deplore the American Revolution just because it caused great hardship to those who fled from its wrath 170 years ago. It is now generally accepted by even the most die-hard descendants of the United Empire Loyalists that it was a good thing for the world.

It still remains for history to judge how good or bad the Chinese Revolution is, but it is certain that contradictory views are possible and legitimate. It is more than doubtful if the prosperous, colonial Tory, robbed of his possessions, ridden out of town on a rail and forced to carve out a new home in the Canadian wilderness, thought well of the American Revolution. Nor did he appreciate the American attempts to liberate Canada by invading it on four occasions — in 1778, in 1812, and the later Fenian raids. He hated the revolution while others rejoiced in it.

If the rebellion of a few million British colonists, settled on a new continent for only a hundred and seventy years, has had such great repercussions on the Western world, how can we afford to dismiss this revolution of an ancient civilization as a transient thing and cling to the belief that two million dispossessed unfortunates in Formosa can remain the governing force?

However much we may regret it, there is no point in pretending that the Chinese Revolution did not happen. However much we may deplore the present, the past cannot be undone; and, if we are honest, we must admit that the change was overdue.

Let us make provision for a safe home for refugees from revenge, as Great Britain did in Canada for the Tories, and strive to achieve in due course as amicable a settlement as now prevails between the United States and Canada. Sundry attempts to join Formosa to China can be expected. The Chinese Communists probably consider attempts to invade Formosa in as altruistic a light as the American revolutionaries regarded their invasions of Canada.

For us to understand the brutality of the Chinese Revolution, we must understand that, once the bounds of civilized peace are broken, man is a ruthless and bloodthirsty creature, whether he is a Chinese Communist, a French revolutionary, a German Nazi, or a soldier on Sherman's march to the sea.

In these terms one can perhaps understand the Chinese Revolution; but why, one may ask, did it need to go Communist? And since it has done so, why should it not be abhorred?

After a century of trespass and the imposition of extra-territorial rights by Western nations, China welcomed a strong Chinese government. After millennia of despotism and a generation of civil war and invasion, I can imagine that any strong and efficient government that represented the peasants, that was not corrupt, and that produced peace was welcomed by the majority of Chinese. What could a people, 90 per cent of whom were illiterate and most of whom were near starvation, know or care about democracy? The slow, steady evolution of democratic government had no appeal for the Chinese. They emulated the Russians, who had brought themselves from serfdom to *sputniks* in a generation.

Does that mean that Communism will remain unchanged in China or that it is still sufficient for Russia? On the contrary, the Russians, having built the foundation of their new life under Stalin's cruel dictatorship, are now avidly seeking freedom. They are making progress in industry and in education, not because of Communism, but in spite of it. The individual is regaining his position in the state, so that in Russia today the worker and the student are spurred to work by desire for a better life for themselves, not by thought about the good of the state. The Russians, like the French after their revolution, are seeking by slow and devious paths a route to a freer and more civilized society than can possibly be offered by Communism.

It is in the satellite countries of eastern Europe that one sees the full horror of Communism. At least Russian and Chinese Communism was self-inflicted upon societies that were already wretched and degraded. It offered to the poorest classes a chance to rise. It was brutal but it was effective. Corruption, serfdom and misery were so widespread under the czars and emperors that efficient native dictatorships offered a chance for improvement.

In eastern Europe, on the other hand, Communism was imposed upon relatively free and well-educated people by the force of foreign armies. Here was degradation, doubly evil because it came from without, and because it represented a long step backward for people who were already well advanced in their struggle toward representative government. It is tyranny, foreign despotism, degradation and the expansionist tendencies of Communist governments that we resent particularly. Where these things are lacking, as in Yugoslavia, our sense of tolerance is sufficiently well developed to allow us to acquiesce in people choosing their own form of government freely, even if that form seems wrong to us.

During the International Geophysical Year I had for the first time a chance to visit several Communist countries, and what struck me most about them was how greatly they differed one from another, how little like a united bloc they were. The assumption that a common faith is sufficient to hold dissimilar nations together would seem to be disproved by the warring history of Christian Europe.

To think that Communist nations necessarily agree well together is to succumb to their propaganda. At the same time, to underrate their individual potentialities is to neglect the lesson of the *sputniks*. It would be wiser to overestimate their success in our planning for the future.

We must grasp and seriously cogitate upon the idea that the Chinese and Russians are human beings very like ourselves and that the next war has a good chance of destroying most of humanity. At present two great powers, the Americans and the Soviets, are able to destroy one another, and they will soon be joined in the "Nuclear Club" by other nations, including China. We must consider what we can do to avoid the destruction that lies so clear and grim before us. It is quite probable that there is no answer, that man—being constituted as he is—will vanish from the face of the earth as many another animal before him; but it seems worth a try to save ourselves.

One popular belief is that all would be well if the Communists would stop being beastly. Since, to the Communists, the simple solution to the world's uneasy state is for the West to stop being beastly over Formosa and Berlin, this approach obviously leads nowhere. Another suggested solution is that we should set a noble example by starting to disarm ourselves unilaterally. In the present climate of distrust and hatred this seems foolish, and is certainly not in the realm of practical likelihood.

The traditional stand has been to engage in an armament race in which each side fears and distrusts the other and therefore, in the name of defence, builds up a stockpile of arms and increases propaganda. Such actions are regarded as offensive by the adversaries, and so step by step mutual fear, distrust, and imprecations mount until some incident sparks a long and costly war.

Wasteful and disastrous as were wars in the past, they were never so destructive as to prevent recovery by one or both sides within a few years. In North America we have been fortunate that for nearly a century there has been no warfare here. It is, therefore, particularly difficult for us to visualize how terrible an atomic war would be. Since the avoidance of a major war is the only defence known, and since this is so much in all the world's interest, there is perhaps a slight hope that this may be achieved. If so, how is it to be done and who is to take the lead in the search for a solution?

The best hope for a solution seems to lie along the road that the civilized have always pursued — the path of intellect rather than passion. Can we use our minds to recognize that while many of us live in greater luxury than man has ever known before, all of us live in greater danger? Can we recognize that mankind's greatest problems are common problems for all men? Can we overcome our mistrust of and hostility to strangers sufficiently to co-operate with them? Can we see that the more we hold ourselves to be better than others, the more it behooves us to offer leadership? Can we

257

see that although governments and individuals differ, the human race is everywhere much the same?

It seems to me that there is some hope in this direction. There have been too many wars in recent years for any nation to be really spoiling for a fight. The Russians, defeated in 1905 and 1917, invaded in 1919 and 1941, and torn by civil war and purges, have lost scores of millions within living memory. Since meeting Russian scientists, I have had no question in my mind that the common people's clamour for peace is real. The situation is similar in China. However belligerent they may appear, the Chinese Communists have as yet neither the industrial might nor the atomic weapons with which to support a major war, even if they are seeking them. The people in both these countries want to improve their lot, they see the means to do so, and their leaders have promised that they will attain this end.

In the West men are much too comfortable to want to fight. Everywhere men are being educated, which we must believe will have a civilizing and democratic influence. Means of quick communication and transportation exist which should increase understanding and facilitate the exchange of artists, athletes, business men, and scholars, all of whom can demonstrate how essentially similar human beings are.

The leaders of all countries must know that atomic war can only wreck everyone and they must realize that, relieved of the burden of defence, the technical means exist to provide abundance for all and perhaps to stem the overpowering flood of the world's population. This intellectual approach would appear to be the only road to salvation.

In advocating it one naturally thinks of the universities, which traditionally produce the world's intellectuals and are the chief custodians of learning. Education is now so highly regarded and so widely sought that universities could do much by setting an example in intelligent leadership. I was impressed by the zeal, the excellence, the high scholarship, and the breadth of knowledge of Russian and Chinese universities. They are strong and they have a tremendous influence in their own countries. They are reaching out with interest to students from many other lands. If I was received in China with urbane hospitality it was not only because revolution had not obliterated the imprint of three thousand years of civilization, but because I carried the credentials of international science. Unofficial and ensign courier though I was, they were stamped with the same frank that had served Matteo Ricci, Benjamin Franklin, and Sir Humphrey Davy.

Since the problems are pressing and the time for their solution is short, we may well ask if the Western universities are offering as active and appropriate a leadership as their role and the importance of the problems demand.

As is well known, our universities arose with the revival of learning in Europe, and they devoted themselves to the mainstays of that civilizing growth — Greek philosophy, Roman law, and Christian religion.

These studies were chosen not only because they were good, but because they were almost the only ones possible at a time when the Americas were undiscovered, most of Africa was unknown, Cathay was a fable told by Marco Polo, and science was little more than alchemy.

For centuries the main stem of all Western education was the relation of man to his Maker, and the study of man and civilization within the limits of classical belief. There was too much for any man to learn it all, but it was sufficiently limited for many to be acquainted with the whole framework. Indeed a knowledge of the outlines of Western civilization became the test of an educated man. Knowledge of other matters counted for little. In parallel fashion, for two millennia the Chinese recruited their civil servants — the literati — by examination in the classics of Confucius.

Starting with the occasional rebel, such as Roger Bacon and Galileo, who flew in the face of tradition, scientists steadily increased in numbers and activity in the West until their efforts revolutionized life on this earth. By opening up communications they disclosed other civilizations and provided easy means of transport to every part of the earth. This had the indirect effect of destroying the ancient, circumscribed Chinese system of education — and it seems a good question to ask whether it is not time that the Western system was considerably modified also.

Unhappily, the scientists, in their preoccupation with the pursuit of knowledge, have taken little responsibility for the results of their search. They often fail to realize that without great effort there may soon be no civilization within which to pursue their quest. Perhaps for this reason, their work has made less impact on the classical centres of the universities than on the rest of the world. The humanists, the custodians of the classical tradition, still dominate Western universities. Resenting the power of science, they seek to oppose it by pretending that life has not changed; satisfied with the "eternal verities", they are blind to much else in life. Many of them pride themselves on a total lack of scientific knowledge. Is not this attitude as irresponsible as that held by the scientists?

Many platitudes are uttered every day about the wonderful broadening and civilizing effect of the teaching of the humanities; but are they valid? Musing idly on the Trans-Siberian Railway, I often wished that I had had more education in the humanities so that I should have more knowledge of the languages, literature, history, religion and philosophies of the people among whom I travelled, until I reflected that the average course in the humanities would have taught me little of those countries since they lay outside the classical orbit.

Can this preoccupation with one particular past educate the realistic leaders whom the world needs? It is not that I am opposed to classical studies, for they are admirable training; but should not the base be broadened to include other civilizations than our own? In one typical Western university there are thirty professors of classics but only three of East Asiatic studies, one lecturer in Indian studies, and none for most of Africa; there are forty-one professors of the languages of western Europe but only two of Russian.

How can we hope to understand and appreciate other civilizations whose literature and philosophy are unknown to all but a few scholars? Today there is a notable interest in foreign languages in Russian and Chinese universities, and they are better and more extensively taught there than in the West. It seems fair to suggest that our universities are not giving the leadership to the West that they might, for on the one hand the scientists, who are looking at the road ahead, do not want to steer, and on the other the humanists, who do, insist on looking back over their shoulders and on ignoring the other traffic.

What is clearly needed, in place of the present dichotomy in the universities between irresponsible science and narrow humanities, is the advocacy of a broader humanism with a regard for all mankind and a sense of the power and responsibility of science to succour all men. This is the civilizing leaven that the universities should cultivate. When one asks who is to provide the leadership the world so desperately needs, one must conclude that the approach must be intellectual and international, that the universities must accept the responsibility of producing leaders with a universal vision, and that it is the West that must strengthen and continue the leadership it has already shown.

That it is the West that must provide the leadership can be understood immediately we stop to ask ourselves: "Who, at the present time, is following whom?" Are we throwing off our trousers and putting on mandarin coats? Are we growing pigtails? Have we stopped mass production in favour of handicrafts? Who was Marx, and where did he work? It is Western ways that the world wants. The common pride and perversity of human nature prevents the Chinese from

261

admitting this, but it remains nonetheless true that they have abandoned many of the trappings of a civilization three thousand years old to adopt many of the new ideas of the West, as the Russians did before them. We should recognize this and do our best to be worthy of the leadership thrust upon us.

The easy course is to say that the Communist nations are in the wrong, therefore it is up to them to correct their errors. The more mature view is that we must continue to lead because we believe the Communists are wrong and because they are in reality endeavouring to copy our ways. If the West is right, then the West must lead.

If the Western nations can find a course of dispassionate moderation, tolerance, and justice and lead the way along it, the world will be indebted to them for discovering the way to a better life. If they fail, any survivors of the holocaust will be lucky to find an unscorched patch of jungle from which to start the long climb back to civilization.

Index

264

265

266

267

269

271

CPSIA information can be obtained
at www.ICGtesting.com
Printed in the USA
BVHW031433300119
539055BV00002B/16/P